Defects and Properties of Semiconductors:
Defect Engineering

Defects and Properties of Semiconductors: Defect Engineering

Edited by
J. Chikawa, K. Sumino, and K. Wada

KTK Scientific Publishers/Tokyo

D. Reidel Publishing Company
A MEMBER OF THE KLUWER ACADEMIC PUBLISHERS GROUP

Dordrecht / Boston / Lancaster / Tokyo

Library of Congress Cataloging-in-Publication Data

Defects and properties of semiconductors.

(Advances in solid state technology)
"Papers presented at the Symposium on "Defects and Qualities of Semiconductors" which was held in Tokyo on May 17-18, 1984 under the sponsorship of the Society of Non-Traditional Technology"--Pref.
  1. Semiconductors--Defects--Congresses.
I. Chikawa, J. (Junichi), 1930–   . II. Sumino,
K. (Kōji), 1931–   . III. Wada, K. (Kazumi),
1950–   . IV. Symposium on "Defects and Qualities
of Semiconductors" (1984: Tokyo, Japan) V. Society
of Non-Traditional Technology (Japan). VI. Series.
QC611.6.D4D4184   1987   621.3815′2   86-21992
ISBN 90-277-2352-4

Published by KTK Scientific Publishers (KTK),
307 Shibuyadai-haim, 4-17 Sakuragaoka-cho, Shibuya-ku, Tokyo 150, Japan,
in co-publication with D. Reidel Publishing Company, Dordrecht, Holland

Sold and distributed in the U.S.A. and Canada
by Kluwer Academic Publishers,
101 Philip Drive, Assinippi Park, Norwell, MA 02061, U.S.A.
in Japan by KTK Scientific Publishers (KTK),
307 Shibuyadai-haim, 4-17 Sakuragaoka-cho, Shibuya-ku, Tokyo 150, Japan

In all other countries, sold and distributed
by Kluwer Academic Publishers Group,
P.O. Box 322, 3300 AH Dordrecht, Holland

All Rights Reserved

Copyright © 1987 by KTK Scientific Publishers (KTK)

No part of the material protected by this copyright notice may be reproduced or utilized in any form or by any means, electronic or mechanical, including photocopying, recording or by any informational storage and retrieval system, without written permission from the copyright owner.

Printed in Japan

Preface

This volume contains nearly all of the papers presented at the Symposium on "Defects and Qualities of Semiconductors" which was held in Tokyo on May 17–18, 1984, under the sponsorship of the SOCIETY OF NON-TRADITIONAL TECHNOLOGY. The Symposium was organized by the promoting committee of the research project "Quality Developement of Semiconductors by Utilization of Crystal Defects" sponsored by the Science and Technology Agency of Japan.

Defect study in semiconductor engineering started originally with seeking methods how to suppress generation of harmful defects during device processing in order to achieve a high yield of device fabrication. Recently, a new trend has appeared in which crystal defects are positively utilized to improve the device performance and reliability. A typical example is the intrinsic gettering technique for Czochralski silicon. Thus, a new term "DEFECT ENGINEERING" was born. It is becoming more important to control density and distribution of defects than to eliminate all the defects. Very precise and deep knowledge on defects is required to establish such techniques as generation and development of defects desired depending on type of devices and degree of integration. Electrical, optical and mechanical effects of defects should be also understood correctly. Such knowledge is essential even for eliminating defects from some specified device regions. It is the time now to investigate defect properties and defect kinetics in an energetic way.

From this point of view, all the speakers in this symposium were invited among the most active investigators in the field of defect engineering in Japan. There was a large attendance at the Symposium.

This volume includes the new progress after the symposium as well as current issues discussed in the Symposium and offers the present status and future prospects of defect study of Si and III–V compound semiconductors such as GaAs, InP and GaAlAs. We hope that this book will be a milestone in the "defect engineering".

Finally, we thank the officers of the Office of Materials Science and Technology, the Science and Technology Agency, for their support and encouragement in the planning and execution of the symposium. We are also deeply appreciative of the extensive support by the secretarial staff at the Society of Non-Traditional Technology.

April 1986

J. Chikawa
K. Sumino
K. Wada

CONTENTS

Preface .... J. CHIKAWA, K. SUMINO, and K. WADA   v

PART 1: COMPOUND SEMICONDUCTORS

Dislocations in GaAs Crystals .... K. SUMINO	3
Dislocations in GaAs Crystals Grown by As-Pressure controlled Czochralski Method .... K. TOMIZAWA, K. SASSA, Y. SHIMANUKI, and J. NISHIZAWA	25
Deep Level Photoluminescence in GaAs .... M. TAJIMA	37
Analysis of Nonstoichiometry and Doped Inpurities in GaAs by X-Ray Quasi-Forbidden Reflection (XFR) Method .... I. FUJIMOTO	71
Growth of Dislocation Free InP Single Crystals.......... S. SHINOYAMA	87
InP MISFETs Technology.. T. SUGANO	99
Characterization of Alloy Semiconductors..................... T. KATODA	111
Electrical Properties of DX Center in Selectively Doped AlGaAs/GaAs Heterostructure......................... M. TAKIKAWA and M. OZEKI	133

PART 2: SILICON

Point Defects and Impurities in Silicon Crystals .... J. CHIKAWA	143
The Behavior of Point Defects in Silicon Crystals .... S. MIZUO and H. HIGUCHI	155
Point Defects and Stacking Fault Growth in Silicon .... K. WADA and N. INOUE	169
The Characteristics of Nitrogen in Silicon Crystals......................... .... T. ABE, T. MASUI, H. HARADA, and J. CHIKAWA	185
Oxygen in Silicon .... N. INOUE, K. WADA, and J. OSAKA	197
On the Formation Process of Thermal Donors in Czochralski-Grown Silicon Crystal .... M. SUEZAWA	219
Interaction of Dislocations with Impurities in Silicon........ K. SUMINO	227
Author Index..	261

PART 1: COMPOUND SEMICONDUCTORS

Defects and Properties of Semiconductors: Defect Engineering, edited by J. Chikawa, K. Sumino, and K. Wada, pp. 3–24.
© KTK Scientific Publishers, Tokyo, 1987.

DISLOCATIONS IN GaAs CRYSTALS

Koji SUMINO

The Research Institute for Iron, Steel and Other Metals, Tohoku University, Sendai 980, Japan

Abstract. A review is made of various aspects of dislocations in GaAs crystals such as atomic structure of dislocations, dynamic characteristics of dislocations under stress, interactions of dislocations with impurities, mechanical strength, and electrical and optical activities of deformation-induced defects. Special emphasis is laid on the comparison of dislocation characteristics in GaAs crystals with those in silicon crystals.

1. Introduction

In reviewing the literatures on defect properties of GaAs crystals, one meets many embarrassing facts concerning unbelievably large disagreements among the data published by different research groups. An extraordinarily large scatter is seen even in the data of an individual research group depending on the specimens used. Further, in spite of a huge amount of papers that have been published on various aspects of defect properties, they are mostly fragmentary and seem to have no correlation to each other. These show how sensitively defect properties are affected by a variety of material factors.

A more or less similar situation was observed on silicon crystals about a decade ago. At present, far more reliable information is available on defect properties of silicon crystals. It owes mostly to the development of growth technique of highly perfect crystals of high-purity silicon and also to the establishment of the method that eliminates the effects of thermal history of a silicon crystal at the time of crystal growth.

The maximum attainable purity of GaAs material at present is lower than that in silicon material. The deviation from the stoichiometry composition also accompanies to compound semiconductors. These complicate any defect-related phenomenon occurring in a GaAs crystal in comparison with a silicon crystal. Consequently, in the investigation on dislocations in GaAs crystals, their interaction with point defects of various kinds is one of the most important subjects to be clarified. In the presence of such complicating factors the information that has been obtained on silicon crystals may play the role of an important guidepost for the study of dislocations in GaAs crystals. From such viewpoint, several aspects are reviewed concerning the characteristics of dislocations in GaAs crystals in the following in comparison with those in silicon crystals.

2. Atomic Structure of Dislocations

Dislocations that are active in a GaAs crystal under stress at elevated temperature have the Burgers vectors of $(a/2) <1\bar{1}0>$ and glide along the $\{111\}$ planes. These Burgers vectors and glide planes are in common with those in a silicon crystal. In a GaAs crystal (111) atomic planes to be occupied only by Ga atoms and those only by As atoms are stacked alternatively, repeating a wide and a narrow interatomic planes also alternatively as shown in Fig. 1. A (111) atomic plane is connected to the neighboring (111) atomic plane by one interatomic bond per atom across the wide interatomic plane, and by three bonds per atom across the narrow interatomic plane. Since the binding energy between neighboring (111) atomic planes is supposed to be larger at the narrow interatomic plane than at the wide one, it has been believed for long time that slip of the crystal takes place along the wide (111) interatomic planes and, consequently, glide dislocations are also defined at such wide interatomic planes. Dislocations in a crystal with sphalerite structure prefer to lie along $<110>$ directions for the energetic reason. So, most commonly observed dislocations in such a crystal are of 60° type or of screw type. Schematic pictures of the atomic arrangements around a 60° and a screw dislocations in this model are shown in Fig. 2. Dislocations in this model are called shuffle set

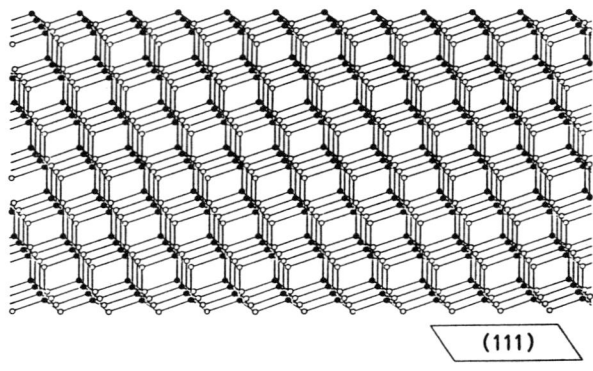

Fig. 1. Schematic picture of the atomic arrangement in a GaAs crystal. Open and full circles distinguish atoms of different kind.

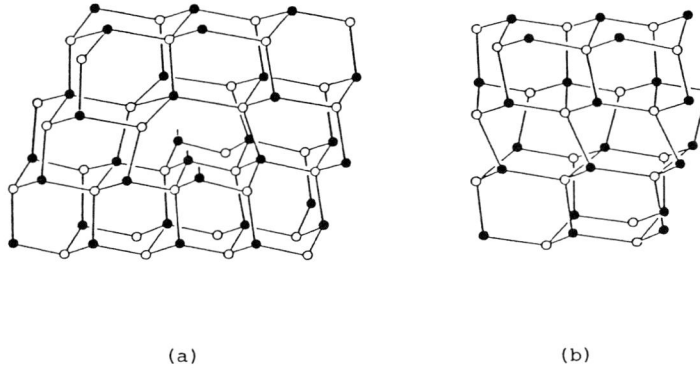

(a)　　　　　　　　　　　　　(b)

Fig. 2. Schematic pictures of the atomic arrangements around shuffle set dislocations in a GaAs crystal. (a) a 60° dislocations, (b) a screw dislocation.

dislocations.[1] A dislocation with an edge component has a row of Ga atoms or As atoms at its core, namely, at the edge of the extra half atomic plane it accompanies, depending on the sign of the Burgers vector. The dislocations with Ga atoms at the core were termed Ga dislocations or α dislocations and those with As atoms at the core As dislocations or β dislocations. Customarily the term α or β dislocation is used to mean a 60° dislocation of plus or minus sign in the sphalerite structure. If white and black circles in Fig. 2 mean Ga and As atoms, respectively, introduction of Ga dislocations causes the bending of the crystal concave downwards and that of As dislocations the bending concave upwards. The former type of bending is termed Ga bending or α bending and the latter type of bending As bending or β bending.

Recent observations on dislocations in GaAs crystals by means of high resolution transmission electron microscopy have revealed that glide dislocations in GaAs crystals are dissociated into extended dislocations accompanying narrow strips of stacking faults between partial dislocations.[2-5] This situation is the same as that in silicon crystals.[6,7] Figure 3 shows a high resolution transmission electron micrograph of a 60° dislocation in a plastically deformed GaAs crystal.[4] A stacking fault in the sphalerite structure is defined at the narrow interatomic plane mentioned above. Consequently, an extended dislocation in a GaAs crystal also exists at such an interatomic plane that has a higher density of interatomic bonds. In silicon crystals it has been confirmed by direct observations that glide dislocations are in the dissociated state both at the time of the generation and during glide motion.[8] It may be natural to suppose that the situation is the same in GaAs crystals. Thus, actual atomic arrangements around glide 60° and screw dislocations that are dominant in a GaAs crystal may be those schematically

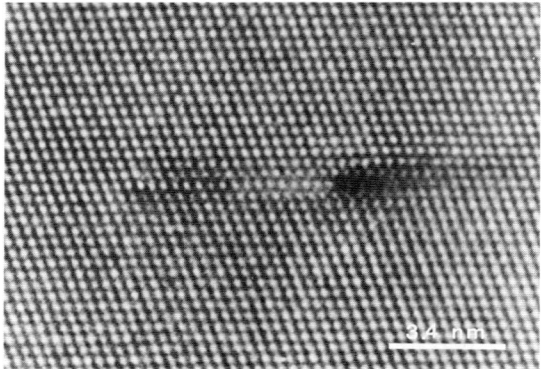

Fig. 3. An end-on view of a 60° dislocation in a GaAs crystal by high resolution electron microscopy. (Sato, Hiraga and Sumino[4]).

shown in Fig. 4. Dislocations of this type are called glide set dislocations.[1] Contrary to the expectation so far prevalent, glide set dislocations have lower Peierls potential than shuffle set dislocations. It is to be noted that core atoms of the glide set dislocations introduced in excess by Ga bending are As atoms and those introduced in excess by As bending are Ga atoms; the situation is just contrary to the case of shuffle set dislocations. At present the terminology of α and β dislocations refers to those of the shuffle set even if the dislocations concerned are in reality glide set dislocations. Consequently, a Ga (α) dislocation has a row of As atoms at the end of its extra half atomic plane and an As (β) dislocations a row of Ga atoms. As seen from Fig. 4, a 60° dislocation consists of a 90° α partial and a 30° α partial or of a 90° β partial and a 30° β partial, both connected by a strip of stacking fault. A screw dislocation consists of a 30° α partial and a 30° β partial.

3. Dislocation Mobility and Impurity Effect

3.1 Dislocation velocity in nondoped GaAs crystals

Grown-in dislocations in GaAs crystals are usually found to be immobile under applied stress at elevated temperature. In situ observations of motion of dislocations freshly introduced into GaAs crystals by means of transmission electron microscopy[9] have revealed that α dislocations move under applied stress at elevated temperature in an extremely jerky way on a scale of 1μm. They are frequently trapped by discrete obstacles of which nature has not yet been identified and move very fast over the distance from one obstacle to the next. Contrarily, β dislocations are observed to move in a viscous manner at velocities much lower than those of α dislocations.

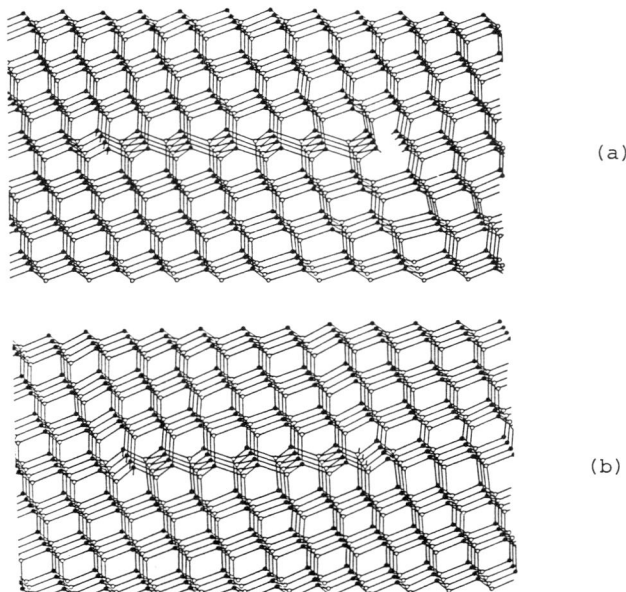

Fig. 4. Schematic pictures of the atomic arrangements around glide set dislocations in a GaAs crystal. (a) a 60° dislocation, (b) a screw dislocation.

The velocities of dislocations in boat-grown GaAs crystals have been measured as a function of the stress and the temperature by several groups by means of etch pit technique.[10-13] The disagreements among the published data are remarkable and velocities measured by different groups under the same stress at the same temperature sometimes differ by two orders of magnitude. Figure 5 shows velocities of various types of dislocations in nondoped crystals under a shear stress of 10 MN/m^2 plotted against the reciprocal temperature. Mark$\alpha_i(i=1-4)$ shows the velocity of α dislocations, the suffix i labelling the research group that made the measurements. Similarly, $\beta_i(i=1-4)$ and $s_i(i=1,2)$ show the velocities of β and screw dislocations, respectively. In spite of large discrepancies among the data of different groups, it is clearly seen that the velocity of α dislocations is higher than that of β dislocations by two or three orders of magnitude. The velocity of screw dislocations seems to be closer to that of β dislocations rather than to that of α dislocations. The velocities of 60° and screw dislocations in a high purity silicon crystal[14] are also shown in the figure for the comparison sake. No detectable motion of dislocations is observed for silicon crystals even under a high stress in the temperature range where velocity measurements have so far been conducted for GaAs crystals. The dislocation mobility in a silicon crystal in the

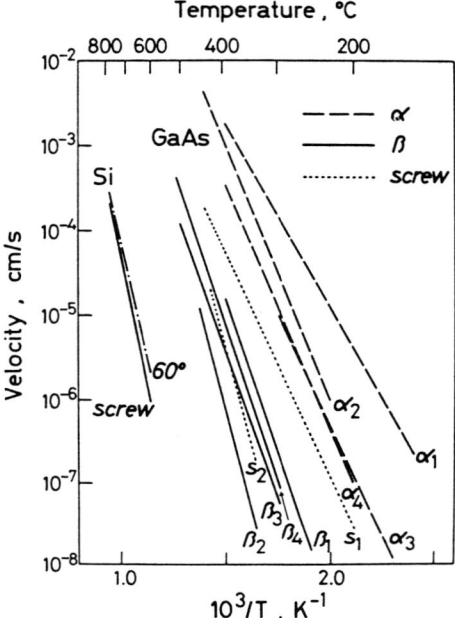

Fig. 5. Dislocation velocities under a shear stress of 10 MN/m² in nondoped GaAs and high purity silicon crystals as dependent on the reciprocal temperature 1/T. Marks α, β and s mean velocities of α, β and screw dislocations, respectively. Suffixes 1, 2, 3 and 4 show that the data are taken from references 10, 11, 12 and 13, respectively.

temperature range 700–900°C is roughly equal to that of the slower type dislocations in a GaAs crystal in the temperature range 400–550°C. It is thought that the Peierls potential in a GaAs crystal is lower than that in a silicon crystal owing to the smaller strength of covalent bonding.

As has already been seen, a screw dislocations consists of two different kinds of 30° partials. Recently, screw dislocations of opposite signs are found to have different mobilities.[9] The screw dislocation of which leading partial is α is observed to move at a velocity faster than that of which leading partial is β by a factor about two at 350°C. This means that the mobility of a 30° partial of any kind depends on whether it moves so as to generate or eliminate the stacking fault.

The dislocation velocity v at a constant temperature is often expressed as a function of the stress τ in such a way as $v \propto \tau^m$. The magnitudes of m determined for various types of dislocations in GaAs crystals have been reported to be in the range of 1.4–2.0. This magnitude of m is slightly higher than that in silicon crystals that is exactly unity.[14] However, the reliability of the measurements of dislocation velocities for GaAs crystals so far conducted seems to be lower than that for silicon crystals. An interesting fact is observed if one extrapolates the data of dislocation mobility to the higher and lower temperature ranges where no measurement is available yet. Dislocation mobilities in GaAs and silicon crystals are of comparable order of magnitude at temperatures above 1000°C. With decrease in the temperature, the mobility decreases far more rapidly in silicon crystals than in GaAs crystals. In the temperature range of 400–600°C the stresses to keep a certain velocity of dislocations differ by six to three orders of magnitude for silicon and GaAs crystals.

3.2 Effect of impurities on dislocation mobility

Influence of doping of p and n type impurities on the dislocation mobility in GaAs crystals has been investigated by several groups.[10-12] The results reported are summarized in Fig. 6. Doping of p type impurities such as Zn

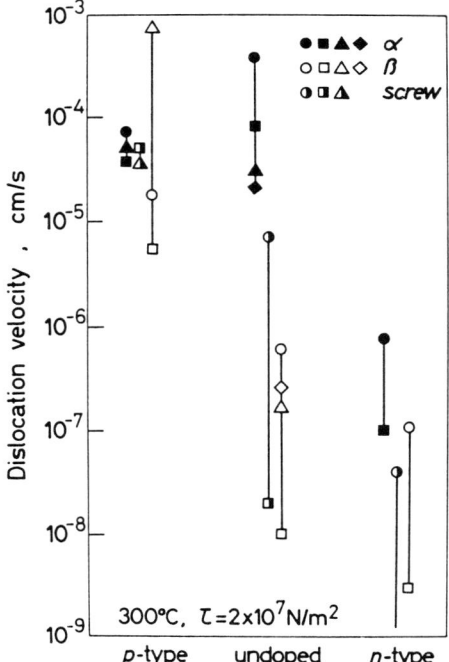

Fig. 6. Effects of doping of p and n type impurities on the dislocation velocities under a shear stress of 20MN/m^2 at 300°C in GaAs crystals. Concentrations of p and n type impurities are in the range of 10^{18}–10^{19} atoms/cm^3 while nondoped crystals have electron concentrations of about 10^{17}cm^{-3}. The same marks are the data taken from the same reference.

leads to an increase in the mobilities of β and screw dislocations while it does not alter the mobility of α dislocations from those in nondoped crystals. Though the magnitudes of the increment in the mobility show large scatter among the data of different research groups, ranging from one to three orders of magnitude, the difference in the magnitudes of the mobilities among different types of dislocations decreases by p doping, and sometimes the reversal in the relative magnitude is observed. No appreciable change is observed in the dislocation mobility in silicon crystals due to p doping.[14,15]

On the other hand, doping of n type impurities such as Te leads to a decrease in the mobilities of all types of dislocations. Amounts of the decrement in the dislocation velocity reported are in the range of one to three orders of magnitude depending on the type of dislocations and also the research groups that conducted the measurements. The decrement is remarkable especially for α and screw dislocations. This result is contrast to the case of silicon crystals for which n doping brings about an increase in the dislocation velocity.[14-16] The enhanced motion of dislocations due to n doping in a silicon crystal has been discussed in terms of the idea that the formation and/or migration energies of kinds on the dislocations are influenced by the position of the Fermi level of the crystal.[17-19] It may be possible to apply the similar idea to the interpretation of the effect of p doping on the dislocation mobility in GaAs crystals. Opposite effects of n doping seen in GaAs and silicon crystals may imply that the positions of energy levels associated with kinks are very different in the two materials.

The lowering of dislocation mobility due to n doping in GaAs crystals can be interpreted with a quite different idea. It is known that n type impurities in a silicon crystal have twofold effects on the activity of dislocations.[14,20] One is the enhancement due to the increase in the dislocation velocity and the other the suppression due to the dislocation pinning. The former is brought about by electrons released from donor impurities and the latter by the electrostatic interaction beetweem ionized donors themselves and some acceptor sites on dislocations.[21] If dislocations in a GaAs crystal have acceptor sites on them as in a silicon crystal, the same kind of electrostatic interaction may also be effective. Dynamic processes of dislocations practically take place in a silicon crystal only in the temperature range above about 600°C and in a lower temperature range in a GaAs crystal. Since the dislocation motion overcoming the interaction with obstacles is usually a thermally activated process, obstruction of impurities or their clusters against the dislocation motion becomes appreciable as the temperature is lowered. Thus, such effects are observed more remarkably in GaAs crystals than in silicon crystals. In other words, the effect of the obstruction of extrinsic origin plays more dominant role in controlling the dislocation motion in a GaAs crystal than in a silicon crystal since the effect of obstruction of intrinsic origin is smaller in it.

3.3 Recombination enhanced motion of dislocations

Numerous reports have been published to show the enhancement of dislocation motion in GaAs crystals and GaAs-based double heterostructure lasers due to irradiation of lights or electron beams and also due to injection of minority carriers. This effect is interpreted qualitatively in such a way that the energy released by recombination of electrons and holes assists the process of dislocation motion overcoming obstacles by some means.

This type of phenomena are observed also in silicon and GaP crystals and, thus, are thought to occur commonly in all types of semiconductor crystals. In the case of silicon crystals, the mobility of dislocations is so low that the recombination enhanced effect is observed only in a relatively high temperature range.[22] Contrarily, the effect is practically important in GaAs crystals with a higher mobility of dislocations since motion and multiplication of dislocations become appreciable even at room temperature due to such enhancement.[23] It is interesting to know the details in the mechanism by which recombination enhanced motion takes place. This is closely related to the rate controlling process in the dislocation motion in semiconductor crystals. Though some discussion is attempted to interpret the effect,[24,25] the exact solution for the mechanism is postponed in the future since no satisfactory theory has yet been established on dislocation motion itself at present.

4. Mechanical Strength

4.1 Dislocation dynamics of plastic deformation

Mechanical behavior of silicon crystals in the temperature range 700–900°C are well established experimentally.[26] Such behavior is almost fully understood in terms of dislocation dynamics[27] using the dislocation mobility known empirically as a function of stress and temperature.

In a GaAs crystal, in which plural types of dislocations having different mobilities are active under stress, deformation is rate-controlled by the motion of dislocations of the type with the lowest mobility. As already seen in Fig. 5, it is at temperatures around 500°C where β or screw dislocations in a nondoped GaAs crystal have the mobility approximately equal to that of 60° or screw dislocations in a high purity silicon crystal at temperatures around 800°C. Thus, comparing the deformation characteristics of a GaAs crystal at temperatures around 500°C with those of a high purity silicon crystal around 800°C, we are able to obtain the idea on whether the dislocation processes controlling deformation or yielding of a GaAs crystal are the same as those of a high purity silicon crystal or not.

Stress-strain curves of nondoped GaAs crystals in compression tests at a constant shear strain rate of $2 \times 10^{-4} \text{s}^{-1}$ in the temperature range 400–500°C are shown in Fig. 7.[28] Figure 8 shows those of high purity silicon crystals in

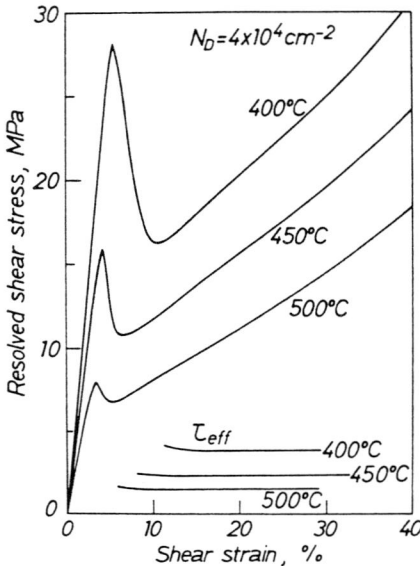

Fig. 7. Stress-strain curves of nondope GaAs crystals in compression tests at a shear strain rate of $2 \times 10^{-4} \text{s}^{-1}$ in the temperature range 400–500°C. The density of dislocations in the crystals prior to compression tests is $4 \times 10^4 \text{cm}^{-2}$. Behavior of the effective stress τ_{eff} against the strain is also shown. (Sumino, Yonenaga, Shibata and Onose[28]).

tensile tests at a constant shear strain rate of $1 \times 10^{-4} \text{s}^{-1}$ in the temperature range 800–950°C.[26] The stress-strain characteristics in the yield region are very similar for GaAs and silicon crystals. They are characterized by marked stress drop after yielding that becomes remarkable drastically as the temperature is lowered. The upper and lower yield stresses increase rapidly as the temperature is lowered. The stress-strain characteristics are observed to be also very sensitive to the strain rate in a quite similar manner in GaAs and silicon crystals.[28] It is commonly observed that slip takes place appreciably in the pre-yield stage and that reloading after unloading in such deformation stage leads the reduction in the magnitude of the upper yield stress. All the above facts show that plastic deformation of a GaAs crystal is controlled by the dislocation processes that are same as those in a high purity silicon crystal, which are now well understood in a quantitative way.[27]

One of remarkable characteristics in a nondope GaAs crystal in as-grown state with a dislocation density of order of 10^4–10^5 cm^{-2} is that the upper yield stress measured under a given deformation condition is very sensitive to the surface condition and tends to show a large scatter. Roughening the surface by abrasion leads to a reduction in the magnitude of the upper yield stress.[28,29]

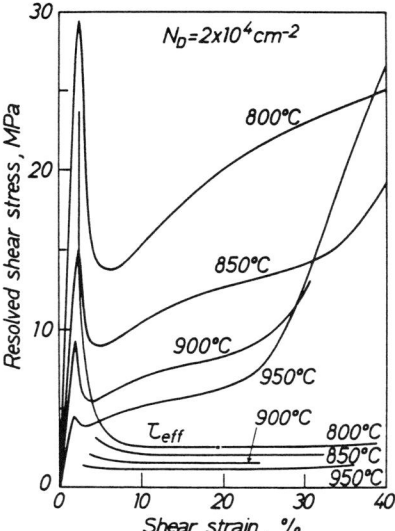

Fig. 8. Stress-strain curves of high purity silicon crystals in tensile tests at a shear strain rate of $1 \times 10^{-4} \text{s}^{-1}$ in the temperature range 800–950°C. The density of dislocations in the crystals prior to tensile tests is $2 \times 10^4 \text{cm}^{-2}$. (Yonenaga and Sumino[26]).

These features agree well with the observations on dislocation-free crystals of high purity silicon.[30]

Contrary to the case of a dislocated crystal of high purity silicon, of which yield characteristics depend very sensitively on the density of dislocations involved in the crystal prior to deformation,[26] those of a GaAs crystal in as-grown state or in well annealed state are little influenced by the density of dislocations involved. Figure 9 compares the upper yield stress versus the dislocation density relations of nondoped GaAs and high purity silicon crystals.[28] Rapid decrease in the upper yield stress with increase in the dislocation density in high purity silicon crystals is known to be due to that dislocations existing prior to deformation act as effective dislocation sources under relatively low applied stress.[27] The density of moving dislocations giving rise to yielding at the upper yield point is attained at low applied stress if the density of dislocation sources is high. The constancy of the upper yield stress against the dislocation density in GaAs crystals seen in Fig. 9 means that dislocations in a GaAs crystal introduced at the time of crystal growth or those subjected to annealing do not act as dislocation sources in the deformation. Probably some kind of surface flaws act as effective dislocation sources. They are activated under stress lower than that starts aged dislocations in GaAs crystals.

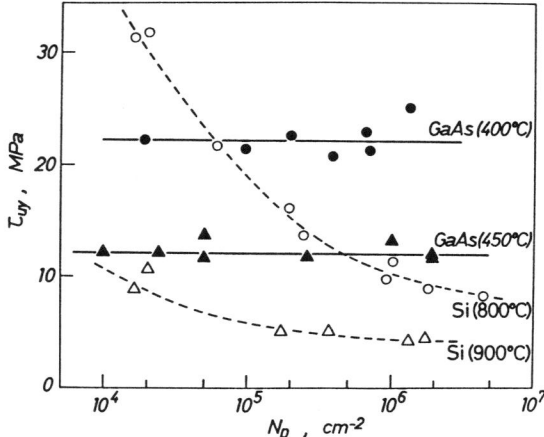

Fig. 9. Dependence of the upper yield stress τ_{uy} on the initial density of dislocations N_D in nondoped GaAs crystals and high purity silicon crystals. The shear strain rates for GaAs and silicon crystals are 2×10^{-4} and $1\times10^{-4} s^{-1}$, respectively. (Sumino, Yonenaga, Shibata and Onose[28]).

4.2 Impurity gettering and dislocation locking

The reason for the inactivity of dislocations in as-grown or well-annealed GaAs crystals is thought to be locking of dislocations by impurities. It is now well established that dislocations in silicon crystals are effectively locked by impurities such as oxygen, nitrogen, phosphorus, etc. at elevated temperature.[31] Such impurities interact strongly with dislocations and gettered by the latter to reduce the total free energy of the system. Hence, once impurity atmosphere has been developed around a dislocation, a high stress is needed to start the dislocation by separating it from the atmosphere.

It has been reported that impurity silicon in a GaAs crystal is effective to immobilize dislocations.[10] Heavy metal impurities such as gold and chromium are also shown to be effectively gettered by dislocations in GaAs crystals.[32,33] Numerous papers have been published to show the facts that electrical or optical properties of the regions close to dislocations in an as-grown GaAs crystal are different from those of the matrix region. This seems to mean that the dislocations have absorbed impurities and/or impurity-point defect complexes from the neighborhood and have improved the purity and perfection of such regions.

Performance of gettering of impurities by dislocations depends on the magnitude of the energy of interaction between a dislocation and an impurity atom and also on the temperature at which the gettering is effectively done. It is a very interesting subject to clarify the mechanism of interaction with dislocations for individual impurities.

4.3 On growing dislocation-free GaAs crystals

Dislocation cell structure is often revealed with X-ray topography, transmission electron microscopy and etch pit techniques in LEC (Liquid Encapsulated Czochralski)-grown GaAs crystals when the density of grown-in dislocations is higher than about 10^4 cm^{-2}. Such dislocation configuration is characteristic of crystals that have been subjected to high temperature deformation. This strongly suggests that main cause for the introduction of dislocations at the time of crystal growth is plastic deformation of the crystal due to thermal stress after it has been crystallized. As has already been mentioned, there seems to be no large difference between the dislocation mobility in GaAs crystals and that in silicon crystals at temperature above 1000°C. Very large difference is seen in low temperature range, where dislocations in GaAs crystals retain high mobility. Thus, it is essential for the growth of a GaAs crystal of a low dislocation density to realize the thermal circumstance that keeps the level of thermal stress very low until the pulled crystal is cooled down to low enough temperature. Controlling the thermal circumstance only around the liquid-solid interface of GaAs to be the same as that in growing a dislocation-free silicon crystal is not enough condition for growing low dislocation density crystals of GaAs.

In addition to reduction of thermal stress in the low temperature region, doping of some special kind of impurities seems to be very effective to grow GaAs crystals of low dislocation densities.[34-38] Impurities of n-type such as Si, S, Se, Te etc. are reported to be effective to reduce the density of grown-in dislocations. As mentioned earlier, this type of impurities have twofold effects on the activity of dislocations; one is reduction in dislocation mobility and the other is locking of dislocations. Probably the latter plays the dominant role in reducing dislocations since it is more effective to suppress the activity of dislocations at elevated temperature under stress. After pulling a GaAs crystal from the melt, dislocations are expected to be generated from some irregularities inside the crystal or on the crystal surface due to concentration or thermal stress. Such generated dislocations will be decelerated and cease moving after the travel over a small distance from the generation center because the stress decreases very rapidly with the distance from the irregularity. At such stage impurities will accumulate on the dislocations. If the thermal circumstance of the crystal is such that the increasing rate of the locking force due to accumulation of impurities is higher than the increasing rate of the force due to thermal stress, the generated dislocations are not able to penetrate into the inside of the crystal and, consequently, do not undergo self-multiplication. Thus, the crystal is kept practically dislocation-free or in the state of low dislocation density. From the above viewpoint, the boat technique is thought to be more favorable than LEC technique in growing GaAs crystals of low dislocation densities because of low thermal stress due to slow cooling and natural doping of impurity Si from quartz boats.

Precipitate particles are often observed on dislocations in LEC-grown GaAs crystals.[39-41] They are of rounded or tetrahedral shape as shown in Fig. 10. Examinations of crystal structure and chemical composition by means of electron diffraction and analytical electron microscopy show that these particles are crystallites of elemental hexagonal arsenic or some crystallites rich in arsenic. By analogy with precipitation behavior of impurity oxygen in silicon crystals, it is suggested that As atoms in excess of stoichiometry composition are accumulated on dislocations and may lock the latter.

Doping of isovalent impurities such as In, Al, N, Sb etc. are also reported to be effective to reduce the dislocation density in LEC-grown GaAs crystals. It is becoming common to grow dislocation-free crystals of GaAs industrially with LEC technique by doping In at concentrations of about 10^{20} atoms/cm^3. If these impurity atoms only replace Ga or As atoms on the lattice sites, no strong interaction is expected between dislocations and the impurities since the lattice distortion around individual impurity atoms is rather small. It may be probable that these impurities that are in excess of stoichiometry composition develop some electrically active complexes. Such complexes may interact electrostatically with electrically active parts on dislocations, resulting in strong locking of dislocations. There may be another function of the impurities that eliminates or inactivates surface irregularities acting as dislocation generation centers. These are the problems to be clarified in future.

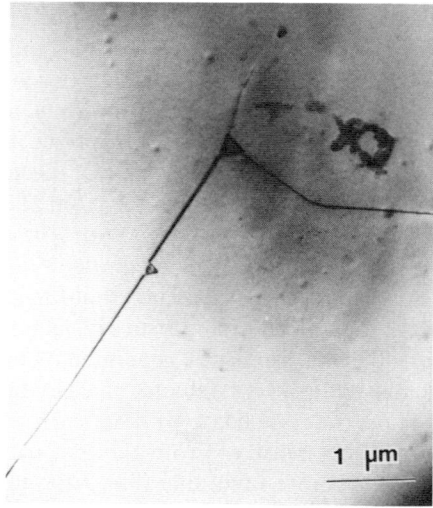

Fig. 10. Precipitate particles developed on a dislocation line in a nondope LEC-grown GaAs crystal. (Sato and Sumino[41]).

5. Electrical and Optical Activities of Defects Introduced by Plastic Deformation

There is a big difficulty in experimental investigation of the electronic states associated with dislocations in semiconductor crystals. Dislocations to be investigated are usually introduced into specimens by means of plastic deformation at elevated temperature. Various types of defects such as clusters of point defects, point defect-impurity complexes, small voids etc. that are all electrically active are simultaneously introduced by deformation together with dislocations. Hence, it is difficult to separate the effect of only dislocations from those of other kind of defects. Investigation of annealing behavior of the effects seems to give an effective clue to extract the effect of dislocations. However, annealing of a deformed crystal at high temperature usually causes segregation of various kinds of impurities on dislocations and may alter the original attribute of dislocations. One must be very careful in interpreting experimental data on plastically deformed semiconductor crystals in terms of the effects of dislocations.

5.1 Studies by the Hall effect

Change in the carrier concentration in GaAs crystals due to plastic deformation has been investigated by few research groups by measurements of the Hall effect.[42-45] The concentration of carriers in n-type material is reduced drastically due to deformation as shown in Fig. 11. This means that a high concentration of acceptors having an energy level in the bottom half of the band gap are introduced due to deformation. The phenomenon is similar to that occurring in silicon crystals, in which main acceptors with an energy level at 0.3-0.4 eV above the valence band are known to be introduced due to deformation.[46] In addition to the acceptors, donors having an energy level roughly at the middle of the band gap seem to be introduced in GaAs crystals due to deformation as supposed from Fig. 11. Such situation is somewhat different from that in silicon crystals.[47] If the amount of deformation exceeds a critical magnitude that is determined by the concentration of shallow donors, an originally n-type silicon crystal changes to a p-type crystal. Such conversion of the conduction type has not yet been found in n-type GaAs crystals. Measurements of the Hall effect of p-type GaAs crystals show that the hole concentration decreases due to deformation. This means that defects acting as donors are introduced into the crystals due to plastic deformation.

5.2 Studies by capacitance techniques

Investigations on the traps for electrons and holes in GaAs crystals associated with deformation-induced defects have been conducted by several research groups[48-52] with the use of capacitance techniques such as deep level

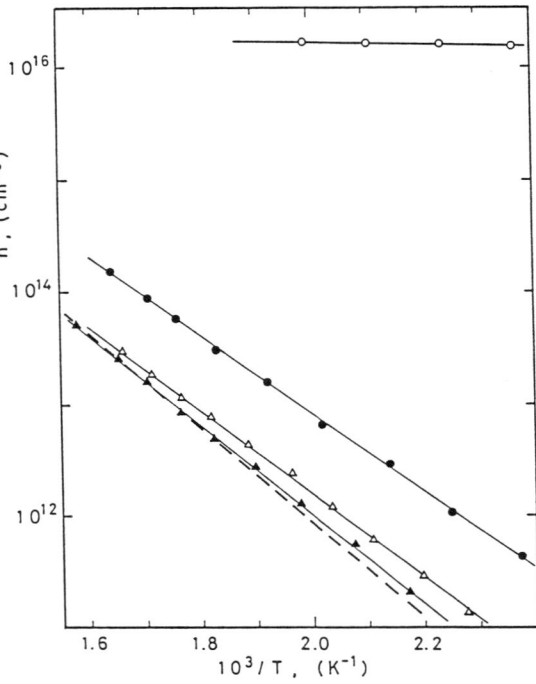

Fig. 11. Relations between the free electron concentration n and the reciprocal temperature $1/T$ before and after deformation in GaAs. Open circles and full circles are the data for n-type conducting crystals before and after deformation of 15%, respectively. Open triangles and full triangles are for semi-insulating crystals before and after deformation of 6.2%. A broken line shows the relation in intrinsic GaAs. (Suezawa and Sumino[44]).

transient spectroscopy (DLTS) and photocapacitance. It is known in the case of silicon that no traps are present in as-grown crystals and that several kinds of traps for electrons and holes are introduced within the band gap due to deformation.[53-55] The concentrations of these traps are found to increase with increasing strain. Most of the traps annihilate almost completely upon annealing deformed crystals at temperatures where no appreciable reduction in the dislocation density takes place. This seems to mean that the deformation-induced traps in silicon crystals are not directly related to dislocations but to debris of dislocation motion and/or some irregularities on the dislocation line such as jogs and kinks.

Contrary to the case of silicon crystals, traps for electrons and holes are already present in as-grown GaAs crystals at concentrations of order of 10^{15} cm^{-3}. The species and the concentrations of such grown-in traps depend strongly on the method by which the crystal is grown. There are large discrepancies among data published by different research groups on the behavior of the trap concentration against the plastic strain. Concerning any specified type of traps one group reports the concentration to increase with strain and another not. One group reports that a new type of traps appear due to deformation while another group insists that the same kind of traps exist in as-grown crystals. The electron traps so far reported have their energy levels at 0.37–0.39 eV, 0.65–0.68 eV, and 0.74–0.84 eV below the conduction band and the hole traps at 0.38–0.4 eV, 0.46 eV, and 0.68 eV above the valence band. Among these traps, electron traps with an energy level at about 0.8 eV below the conduction band are known as EL2. Identification of defects giving rise to EL2 has not yet been accomplished. It may be concluded from the data so far available that plastic deformation of a GaAs crystal does not induce any traps that have their energy levels at the positions quite different from those of grown-in defects.

Recently it is pointed out[52] that there has been a common mistake in the procedure to estimate the trap density of deformed GaAs crystals in the previous works. Namely, in estimating the trap density as a function of the strain the effect of the increase in the concentration of deformation-induced acceptors has not been properly taken into account in the previous works. If this effect is taken into account, the concentration of electron traps such as EL2 is shown to decrease with the strain from the value in the as-grown crystal in the small strain range as seen in Fig. 12.

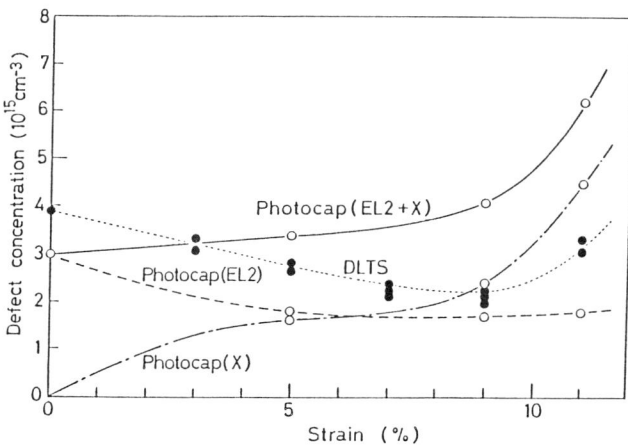

Fig. 12. Variations in the concentration of various defects having the energy levels at the midgap against the plastic strain in GaAs crystals. Full circles are for the defects detected by DLTS and open circles for those detected by photocapacitance. Defects EL2 show photo-quenching effect while defects X not. (Suezawa, Hara and Sumino[52]).

Photocapacitance measurements have revealed the following fact.[52] With the increase in the plastic strain the concentration of the grown-in defects EL2 that show photoquenching effect on irradiation of lights with energies around 1.15 eV decreases gradually while the concentration of the new defects termed X that have the energy level at the same position as that of EL2 and show no photoquenching effect increases steadily as shown in Fig. 12. The behavior of the sum of the concentrations of EL2 and X against the strain does not coincide with that of the defects detected by DLTS. This may imply that the capture cross section of defects X for carriers is smaller than that of EL2 defects.

5.3 Studies by Electron Spin Resonance (ESR)

No ESR-active centers are found in as-grown silicon crystals. Plastic deformation of a silicon crystal at temperature lower than about 800°C induces several kinds of ESR-active centers that are all attributed to clusters of point defects or some irregularities on dislocation lines.[56-58] Contrarily, ESR-active centers are already present in as-grown crystals of GaAs. The situation is similar to the case of the traps for electrons and holes studies by DLTS technique. The concentration of ESR-active centers in a GaAs crystal has been found to increase due to plastic deformation.[44,59,60] The increase in the concentration of such centers is, however, smaller than that in the density of dislocations induced due to deformation by orders of magnitude. Further, they annihilate due to annealing at the temperature where no appreciable decrease in the dislocation density takes place.[61] So, they are thought not to be related to dislocations but to point defects generated during deformation. The ESR signal in a p-type GaAs crystal has been reported to be enhanced by illumination of the light 0.52 eV in the energy and quenched by the lights of 0.75 and 1.0 eV. This behavior has been interpreted on the model that the defects giving rise to the ESR signal are antisite defects in which As atoms occupy the Ga lattice sites.[59] Two mechanisms have been proposed on the creation of antisite defects As_{Ga} due to dislocation motion.[59,62] Such antisite defects As_{Ga} are naturally thought to act as donors of double valence.

Recently, the same kind of ESR signal is found to be observed also in an n-type GaAs crystal.[44] This means that the defects concerned may not be double donors. It is found that the ESR signal in an as-grown crystal shows quenching due to illumination of the light of 0.77 eV and enhancement by the light of energy larger than about 1.1 eV. The amounts of the quenching and the enhancement are almost the same as shown in Fig. 13. Deformation of the crystal leads to an increase in the concentration of the ESR-active centers. The ESR signal shows the same kind of quenching and enhancement effects due to light illumination as in the as-grown crystal. However, the relative magnitude of the enhancement at 1.1 eV in the deformed crystal is much smaller than that of the quenching at 0.77 eV as shown in Fig. 14. This suggests that grown-in

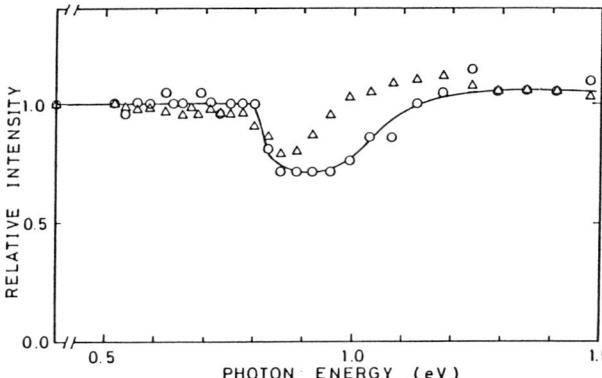

Fig. 13. Variation of the ESR signal intensity against the energy of photons irradiated on an as-grown semi-insulating GaAs crystal. Open circles are for the signal due to defects and open triangle for that due to chromium ions. The signal intensity is normalized with respect to that in the dark. (Suezawa and Sumino[44]).

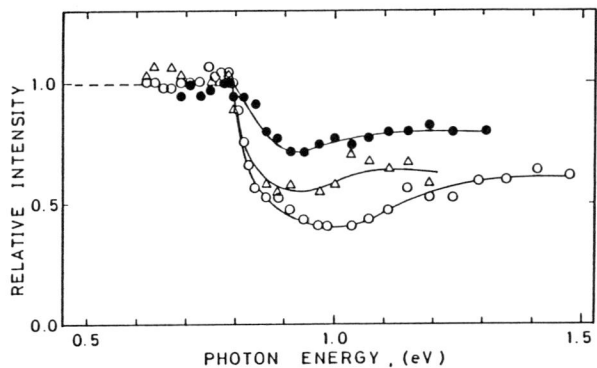

Fig. 14. Variations of the ESR signal intensity against the energy of photons irradiated on specimens. Open triangles are for an originally n-type conducting specimen deformed by a strain of 15%, open circles for a semi-insulating specimen deformed by a strain of 24%, and full circles for a semi-insulating specimen irradiated by electrons of 15 MeV. The signal intensity is normalized with respect to that in the dark. (Suezawa and Sumino[44]).

defects and deformation-induced defects both ESR-active are different from each other. Namely, the former have both quenching and enhancement effects at photon energies of 0.77 and 1.1 eV, respectively, while the latter have only quenching at 0.77 eV. The identification of these two kinds of defects is the subject to be clarified in future.

5.4 Photoluminescence studies

Matrix of a silicon crystal is an indirect semiconductor and shows no photoluminescence. Dislocations in a silicon crystal have been shown to induce sharp photoluminescence lines with energies of 0.812, 0.875, 0.934 and 0.999 eV.[63-66] These lines have been attributed to radiative recombination on dislocations since they are found to survive after annealing that eliminates deformation-induced defects other than dislocations.

In the case of a GaAs crystal that is a direct semiconductor, the intensities of photoluminescences in the energy ranges around the band-gap edge and the mid-gap are all reduced drastically due to plastic deformation, while the shape of the photoluminescence spectrum being almost the same between an as-grown crystal and a deformed crystal.[51,67-69] Cathode-luminescence studies have clearly shown that radiative recombination process ceases to operate in the neighborhood of dislocations.[70,71] Thus, probably dislocations and/or other type of deformation-induced defects act as strong killers for radiative recombination processes that lead to the emission of lights of the above energy ranges. Recently, dislocations are found to induce a broad photoluminescence line centered around 1.1 eV in a GaAs crystal that has an extremely low light emitting efficiency as shown in Fig. 15.[44]

In conclusion, the author emphasizes his view that the quality of GaAs crystals will certainly be improved in many aspects in near future very quickly through fundamental investigations on defect properties and defect-impurity interactions in this material.

A part of the data of author's group quoted in this review has been taken from the work performed through Special Coordination Funds for Promoting Science and Technology from Science and Technology Agency of Japan.

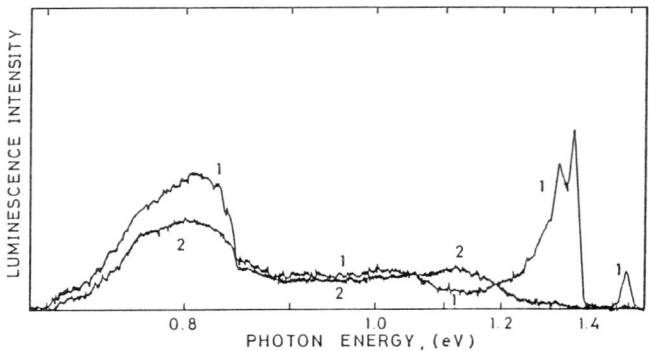

Fig. 15. Photoluminescence spectra of a semi-insulating GaAs specimen before (curve 1) and after (curve 2) deformation of a strain of 24%. (Suezawa and Sumino[44]).

REFERENCES

1) J. P. Hirth and J. Lothe: *Theory of Dislocations* (McGraw-Hill, New York 1968) p. 353.
2) H. Gottschalk, G. Patzer, and H. Alexander: Phys. Stat. Sol. (a) **45** (1978) 207.
3) A. M. Gomez and P. B. Hirsch: Phil. Mag. **A38** (1978) 733.
4) M. Sato, K. Hiraga, and K. Sumino: unpublished work.
5) M. Tanaka and B. Jouffrey: Phil. Mag. **A50** (1984) 733.
6) I. L. Ray and D. J. H. Cockayne: Phil. Mag. **22** (1970) 853.
7) M. Sato, K. Hiraga and K. Sumino: Jpn. J. Appl. Phys. **19** (1980) L155.
8) M. Sato and K. Sumino: Phys. Stat. Sol. (a) **55** (1979) 297.
9) M. Sato, M. Takebe, and K. Sumino: *Proc. Yamada Conf. IX on Dislocations in Solids*, ed. H. Suzuki, T. Ninomiya, K. Sumino, and S. Takeuchi (Univ. Tokyo Press, Tokyo, 1985) p. 429.
10) V. B. Osvenskii, L. P. Kholodnyi, and M. G. Mil'vidskii: *Sov. Phys. -Solid State* **15** (1973) 661.
11) H. Steinhardt and P. Haasen: Phys. Stat. Sol. (a) **49** (1978) 93.
12) S. K. Choi, M. Mihara, and T. Ninomiya: Jpn. J. Appl. Phys. **16** (1977) 737.
13) S. A. Erofeeva and Yu. A. Osipyan: Sov. Phys. -Solid State **15** (1973) 538.
14) M. Imai and K. Sumino: Phil. Mag. **A47** (1983) 599.
15) A. George and G. Champier: Phys. Stat. Sol. (a) **53** (1979) 529.
16) V. N. Erofeev and V. I. Nikitenko: Sov. Phys. -Solid State **13** (1971) 116.
17) H. L. Frisch and J. R. Patel: Phys. Rev. Lett. **18** (1967) 833.
18) P. B. Hirsch: J. de Physique, Colloque C6 **40** (1979) C6-117.
19) R. Jones: Phil. Mag. **B42** (1980) 213.
20) K. Sumino and M. Imai: Phil. Mag. **A47** (1983) 753.
21) K. Sumino: J. de Physique, Colloque C4 **44** (1983) C4-195.
22) K. H. Küsters and H. Alexander: Physica **116B** (1983) 593.
23) K. Maeda, M. Sato, A. Kubo, and S. Takeuchi: J. Appl. Phys. **54** (1983) 161.
24) H. Sumi: Physica **117 & 118B** (1983) 197.
25) K. Maeda: *Proc. Yamada Conf. IX on Dislocations in Solids*, ed. H. Suzuki, T. Ninomiya, K. Sumino, and S. Takeuchi (Univ. Tokyo Press, Tokyo 1985) p. 425.
26) I. Yonenaga and K. Sumino: Phys. Stat. Sol. (a) **50** (1978) 685.
27) M. Suezawa, K. Sumino, and I. Yonenaga: Phys. Stat. Sol. (a) **51** (1979) 217.
28) K. Sumino, I. Yonenaga, H. Shibata, and U. Onose: *Proc. 27th Meeting, 145 Committee of JSPS* (1985) p. 91.
29) D. Laister and G. M. Jenkins: J. Mater. Sci. **8** (1973) 1218.
30) K. Sumino and I. Yonenaga: Jpn. J. Appl. Phys. **20** (1981) L685.
31) K. Sumino: *Defects in Semiconductors II*, ed. S. Mahajan and J. W. Corbett (North-Holland, New York 1983) p. 307.
32) T. J. Magee, J. Peng, J. D. Hong, W. Katz, and C. A. Evans: Phys. Stat. Sol. (a) **55** (1979) 161.
33) T. J. Magee, J. Peng, J. D. Hong, C. A. Evans, and V. R. Deline: Phys. Stat. Sol. (a) **55** (1979) 169.
34) Y. Seki, H. Watanabe, and J. Matsui: J. Appl. Phys. **49** (1978) 822.
35) G. Jacob, J. P. Farges, C. Schemali, M. Duseaux, J. Hallais, W. J. Bartels, and P. J. Roksnoer: J. Cryst. Growth **57** (1982) 245.
36) G. Jacob: J. Cryst. Growth **59** (1982) 669.
37) G. Jacob, M. Duseaux, J. P. Farges, M. M. B. van den Boom, and P. J. Roksnoer: J. Cryst. Growth **61** (1983) 417.
38) Yu. N. Bol'sheva, Yu. A. Grigor'ev, S. P. Grishina, M. G. Mil'vidskii, V. B. Osvenskii, and S. S. Shifrin: Sov. Phys. -Crystallogr. **27** (1982) 433.

39) A. G. Cullis, P. D. Augustus, and D. J. Stirland: J. Appl. Phys. **51** (1980) 2556.
40) T. Kamejima, H. Watanabe, and J. Matsui: *Proc. 27th Meeting, 145 Committee of JSPS* (1985) p. 19.
41) M. Sato and K. Sumino: *Proc. 27th Meeting, 145 Committee of JSPS* (1985) p. 25.
42) H. Nakata and T. Ninomiya: J. Phys. Soc. Japan **47** (1979) 1912.
43) D. Gwinner and R. Labusch: J. de Physique, Colloque C6 **40** (1979) C6-75.
44) M. Suezawa and K. Sumino: Jpn. J. Appl. Phys. **25** (1986) No.4.
45) D. Gerthsen and P. Haasen: Acta Physica Polonica, to be published.
46) H. Ono and K. Sumino: Jpn. J. Appl. Phys. **19** (1980) L629.
47) H. Ono and K. Sumino: J. Appl. Phys. **54** (1983) 4426.
48) T. Ishida, K. Maeda, and S. Takeuchi: Appl. Phys. **21** (1980) 257.
49) Y. Kadota and K. Chino: Jpn. J. Appl. Phys. **22** (1983) 1563.
50) T. Wosinski, A. Morawski, and T. Figielski: Appl. Phys. **A30** (1983) 233.
51) K. Böhm and B. Fischer: J. Appl. Phys. **50** (1979) 5453.
52) M. Suezawa, A. Hara, and K. Sumino: Acta Physica Polonica, to be published.
53) L. C. Kimerling and J. R. Patel: Appl. Phys. Lett. **34** (1979) 73.
54) V. V. Kveder, Yu. A. Osipyan, W. Schröter, and G. Zoth: Phys. Stat. Sol. (a) **72** (1982) 701.
55) H. Ono and K. Sumino: J. Appl. Phys. **57** (1985) 287.
56) E. Weber and H. Alexander: J. de Physique, Colloque C6, **40** (1979) C6-101.
57) M. Suezawa, K. Sumino, and M. Iwazumi: *Inst. Phys. Conf. Ser. No. 59* (1981) 407.
58) E. Weber and H. Alexander: *Inst. Phys. Conf. Ser. No. 31* (1977) 266.
59) E. R. Weber, H. Ennen, U. Kaufmann, J. Windscheif, and J. Schneider: J. Appl. Phys. **53** (1982) 6140.
60) T. Wosinski: Crystal Research and Technology **16** (1981) 217.
61) E. R. Weber and J. Schneider: Physica **116B** (1983) 398.
62) T. Figielski: Appl. Phys. **A29** (1982) 199.
63) M. Suezawa, Y. Sasaki, Y. Nishina, and K. Sumino: Jpn. J. Appl. Phys. **20** (1981) L537.
64) M. Suezawa and K. Sumino: Phys. Stat. Sol. (a) **78** (1983) 639.
65) M. Suezawa, Y. Sasaki, and K. Sumino: Phys. Stat. Sol. (a) **79** (1983) 173.
66) M. Suezawa and K. Sumino: J. de Physique, Colloque C4 **44** (1983) C4-133.
67) W. Heinke and H. J. Queisser: Phys. Rev. Lett. **33** (1974) 1082.
68) K. Böhm and D. Gwinner: Appl. Phys. **17** (1978) 155.
69) A. V. Bazhenov and L. L. Krasil'nikova: Sov. Phys. -Solid State **23** (1981) 2068.
70) T. Kamejima, Y. Matsumoto, H. Watanabe, and J. Matsui: *NEC Research & Development*, No. 77 (1984) p. 64.
71) T. Sekiguchi and K. Sumino: unpublished work.

DISLOCATIONS IN GaAs CRYSTALS GROWN BY As-PRESSURE CONTROLLED CZOCHRALSKI METHOD

K. Tomizawa,[1] K. Sassa,[1] Y. Shimanuki,[1] and J. Nishizawa[2]

[1] *Perfect Crystal Growth Group, Nishizawa Perfect Crystal Project, Research Development Corporation of Japan, Central Research Institute, Mitsubishi Metal Corporation, 1-297 Kitabukuro, Omiya, Saitama, Japan*
[2] *Nishizawa Perfect Crystal Project, Research Development Corporation of Japan, Semiconductor Research Institute, Kawauchi, Sendai, Japan*

Abstract. A new growth technique of the low dislocation density GaAs crystals is described. The ambient As vapor pressure was controlled by the lowest temperature in a quartz chamber in which a crystal was grown by the Czochralski method. The dislocation density in the undoped crystals ($2'''^{\phi}$, <100>) was as low as 2×10^3 cm^{-2}, which was lower than the case of the commercial LEC method by a factor of ten.

Analyses of impurities in crystals suggested to be possible to obtain undoped semi-insulating GaAs crystals by this growth technique.

It is well known that dislocations in substrates reduce the lifetime of the laser diodes[1] and it has been shown that the dislocation distribution affects the uniformity of the FET threshold voltage in the IC devices.[2] Therefore, numerous attempts have been carried out to develop the growth technique for dislocation free GaAs crystals.

Nishizawa et al.[3] have revealed that the ambient As vapor pressure determines the stoichiometry and therefore the defect density in GaAs crystals. On the basis of an extensive investigation of the liquid phase epitaxy and annealing of GaAs, they reported the optimum As vapor pressure (P_{opt}) for growth of the stoichiometric GaAs crystal as follows;

$$P_{opt} = 2.6 \times 10^6 \exp(-\frac{1.05}{kT_g}) \tag{1}$$

where T_g is the growth temperature and k is Boltzman constant. In the crystal growth from the melt, putting the melting temperature of GaAs, 1238°C, into eq. (1), one obtains the value of P_{opt} as 813 torr. The As vapor pressure is determined by the lowest temperature, T_{As}, in the chamber in which the crystal is grown. By using the relationship between T_{As} and the As vapor pressure,[4] the temperature corresponding to the As vapor pressure of 813 torr is obtained as 612°C. Thus suppression of dislocation density is expected by growing crystals under the condition of $T_{As} \sim 612°C$.

The dislocation density of undoped GaAs crystals grown by the Horizontal Bridgman (HB) method has been reduced to 5×10^3 cm^{-2} or less owing mainly to the facts that the As vapor pressure control is possible in the commercial HB method and the thermal stress at the growing interface is not large. However, in this method, it is difficult to avoid the generation of dislocations at the bottom surface of a crystal where some excessive stress is exerted due to the contact with the boat.

On the other hand, in the case of the liquid encapushlated Czochralski (LEC) method the dislocation density in the undoped crystals is about 2×10^4 cm^{-2} or more, although the crystals are grown without any contact with a crucible. The following facts appear to be the main difficulties in this method for further reduction of the dislocation density.

First, a crystal is subjected to an excessive thermal stress caused by the high temperature gradient just above growing interface. Second, it is more difficult than the case of the HB method to keep the melt composition to be optimum during the growth.

For the reduction of excessive thermal stress to the crystal, it has been proposed to grow crystals by the LEC method by elevating the ambient temperature.[5] However, it has also been shown that higher ambient temperature tend to generate Ga droplets on the surface of the crystal due to As escape.[6]

Thus, growth of GaAs crystals by the Czochralski method (without liquid encapsulant) under a controlled As vapor pressure will lead to significant reduction of the dislocation density through the combined effects of improvement of the melt composition and reduction of thermal stress without formation of Ga droplets.

Steinman et al.[7] and Leung et al.[8] have shown that low dislocation density crystals can be grown by the Czochralski method under an As atmosphere. However, because of the complexity and limitation of repeated use of their apparatus, their techniques have not been applied for industrial use.

In this study, a new apparatus for the Czochralski method is proposed for growth of large crystals under a controlled As vapor pressure. It is shown that the low dislocation density GaAs is grown by this apparatus and that the variation in crystal diameter affects the dislocation density. Possibility of growth of undoped high resistivity GaAs by this apparatus is also suggested.

Apparatus and Experimental Procedures

A schematic diagram of the crystal puller is shown in Fig. 1. Crystal growth is carried out under a controlled As vapor pressure in a quartz

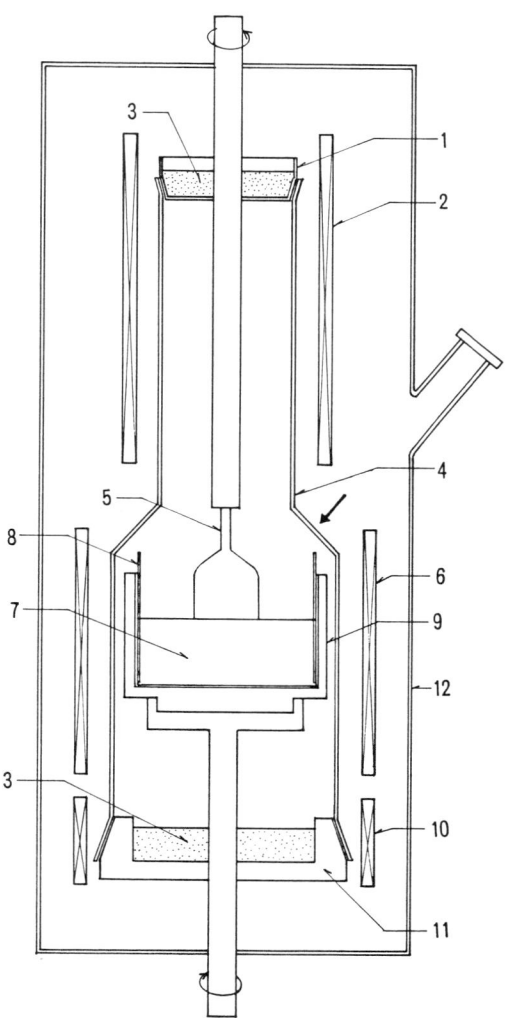

Fig. 1. Schematic diagram of the crystal puller. 1: Upper lid. 2: Heater A. 3: B_2O_3. 4: Quartz chamber. 5: Seed. 6: Heater B. 7: GaAs. 8: PBN crucible. 9: Carbon susceptor. 10: Heater C. 11: Lower lid. 12: Metal chamber.

chamber. The quartz chamber is 60 cm high and 10.8 cm in inner diameter (id), and the pyrolytic boron nitride (PBN) crucible is 8.4 cm in id.

In order to grow crystals under a controlled As vapor pressure the following requirements should be fulfilled. First, the temperature within the quartz chamber must be precisely controlled. The As vapor pressure is controlled by the temperature at the top portion of the quartz chamber and the temperature of the other parts must be maintained above this temperature to avoid As deposition onto the chamber wall. Furthermore, the portion indicated by the arrow in Fig. 1. must be kept transparent during growth in order to observe the growing crystal; the temperature at this portion must be such as to avoid GaAs deposition. These requirements are met by controlling temperatures of three separate carbon heaters shown in Fig. 1.

Second, an effective seal is required between the chamber and the pull rod as well as the crucible supporting rod. As a sealing material, molten B_2O_3 in Mo lids is used.

The third requirement is as follows. After the growth of the crystal, the crystal and the crucible must be taken out of the quartz chamber without fracture of the chamber and the crucible. To meet this, taper joints between the quartz chamber and Mo lids are used.

The growth operation is made as follows. First, after the metal chamber is evacuated, the heaters A and C are heated to melt the B_2O_3 and the upper and lower lids are sealed by the taper joints. Then, the heater B is heated to melt GaAs. During the melting process, N_2 gas is introduced into the metal chamber in order to balance the pressure outside of the quartz chamber against the As vapor pressure within the quartz chamber. After melting the GaAs, the temperature of the heater A is so controlled that the desired As vapor pressure within the quartz chamber is realized. After these processes, a seed is lowered, and a crystal is pulled by the ordinary Czochralski method.

For the growth of each crystal, presynthesized GaAs (600 g) and sufficient As to give a required As vapor pressure over the GaAs melt were charged in a PBN crucible. Two crystals No. 1 and 2 were pulled under the As vapor pressure controlling temperature of 616°C. These crystals were grown without and with B_2O_3 on the GaAs melt to investigate the effects of B_2O_3 on impurity concentration. The water content in B_2O_3 used was 100 ppm. A LEC crystal was prepared to compare the surface feature and dislocation distribution with crystals 1 and 2. A (110) slices, 400~500 μm in thickness was cut from each crystal longitudinally to the growth axis to examine the dislocation distribution by X-ray topography; a Cu X-ray target was used.

For measurement of etch pit density (EPD), (100) wafers were cut from crystal 2 perpendicularly to the growth axis at an interval of 5 mm. These wafers were polished and pre-etched by $H_2SO_4:H_2O_2:H_2O$ for removing the mechanical damage and then etched by molten KOH for revealing the dislocation pits. Measurement of EPD was carried out on wafers along

<110> at an interval of 2 mm.

These three crystals were evaluated by Hall measurement, secondary ion mass spectrometry (SIMS), and localized vibrational modes absorption (LVM). For Hall measurement and SIMS, (100) wafers were cut from the shoulders of three crystals. On the other hand, (110) slices, 5 mm in thickness were cut for the measurement of the C concentration by the LVM method.

Results and Discussion

The crystal grown under an As atmosphere is shown in Fig. 2. For comparison, a crystal by the LEC method is shown in Fig. 3. It should be noted that the surface of the crystal grown under an As atmosphere exhibits metallic luster owing to suppression of the As vaporization from the crystal surface.

X-ray topographs of three crystals shown in Figures 4, 5 and 6, also indicate the difference in the soundness of surface of these crystals. In the case of crystals grown under an As atmosphere (Figs. 4 and 5), surface is smooth and no Ga droplet is found at the periphery. On the other hand, in the case of the LEC crystal (Fig. 6), the surface is rough and many Ga droplets are found

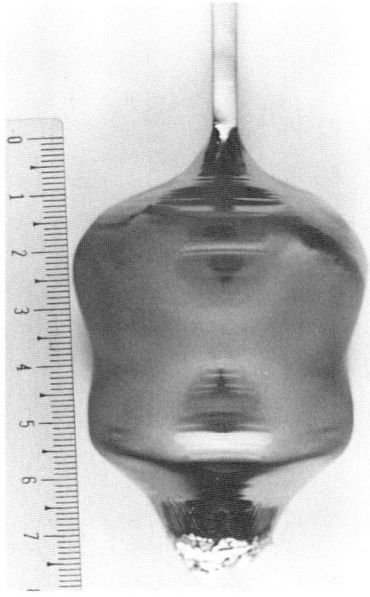

Fig. 2. GaAs crystal grown by As-pressure controlled Czochralski method. The crystal surface exhibits metallic luster.

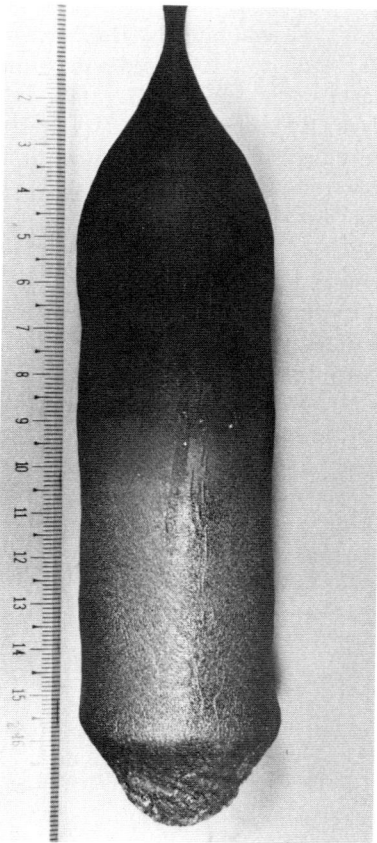

Fig. 3. GaAs crystal grown by the LEC method. The crystal surface is rough.

at the periphery. As for the dislocation distribution, U shaped dislocation distributions are seen in Figs. 4 and 5, whereas a W shaped distribution and coarse slip lines are seen in Fig. 6. It is interesting to note that, in the case of the crystal shown in Fig. 4, some rapid increase of crystal diameter has occurred from the seed end to the shoulder, and at this region dislocation density also increases rapidly. The crystal in Fig. 5 shows gradual increase of crystal diameter. In this case, no rapid increase of dislocation density has occurred. Thus it appears that avoiding the rapid increase of the crystal diameter is an important condition to reduce the dislocation density.

At the periphery of the lower part of the crystal 2 shown in Fig. 5, coarse slip lines are observed. These slip lines seem to generate during the cooling

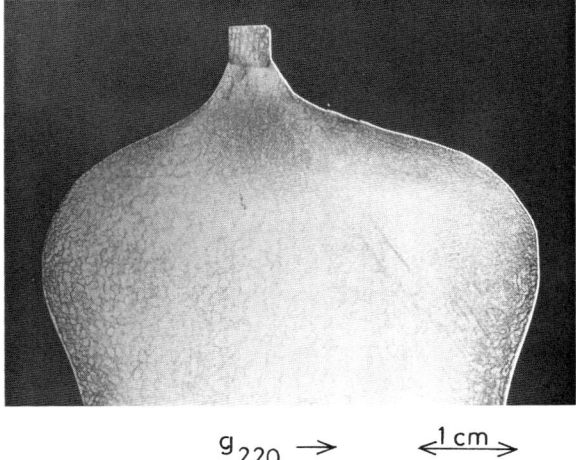

$g_{220} \rightarrow$ $\xleftarrow{1\,cm}$

Fig. 4. X-ray topograph of a longitudinally cut slice of crystal 1 pulled under an As pressure controlling temperature of 616°C.

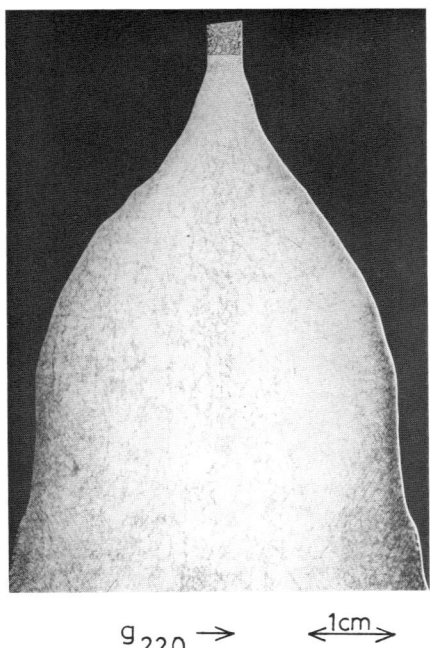

$g_{220} \rightarrow$ $\xleftarrow{1cm}$

Fig. 5. X-ray topograph of a longitudinally cut slice of crystal 2 pulled under an As pressure controlling temperature of 616°C with B_2O_3 on the melt.

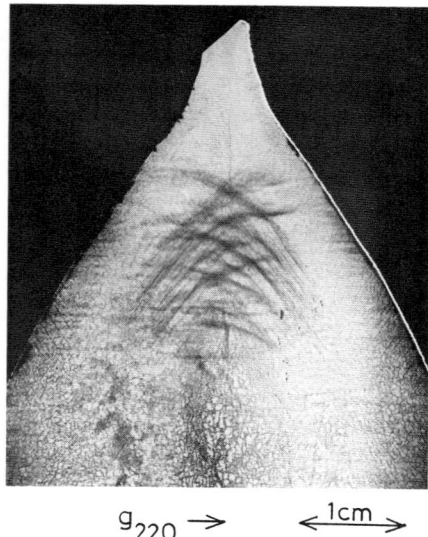

$g_{220} \rightarrow$ $\leftarrow$ 1cm $\rightarrow$

Fig. 6. X-ray topograph of a longitudinally cut slice of crystal 3 pulled under an argon atmosphere. Many Ga droplets are found at the periphery due to the As escape from the surface.

process after the growth since the cooling rate was rather high in this case. It is expected that these slip lines are suppressed by keeping the cooling rate slower.

In Fig. 7, the dislocation density distribution of the crystal 2 evaluated by the etch pit counting is shown. Except for the periphery, the dislocation density is as low as 2,000 cm^{-2}; dislocation free regions are seen. In Fig. 8, the dislocation free region is shown. Figure 9 shows the dislocation density distribution along the growth axis at an interval of 5 mm. The dislocation density is low and the distribution is almost constant in comparison with the case of LEC crystals.

Possible explanations for the pronounced reduction of the dislocation density achieved by this study are as follows. First, improvement of the composition of GaAs melt due to the As vapor pressure led to suppression of point defect formation and in turn dislocation formation. It is possible that dislocation loops are formed by aggregation of point deffects.[9] In addition, climb motion of dislocations, which is assisted by point defects,[9] will reduce the back-stress acting on dislocation sources and lead to generation of new dislocations. Second, since As vaporization from the crystal surface was suppressed by the As atmosphere, the growth of crystals under a low temperature gradient along the pulling axis was possible without formation of

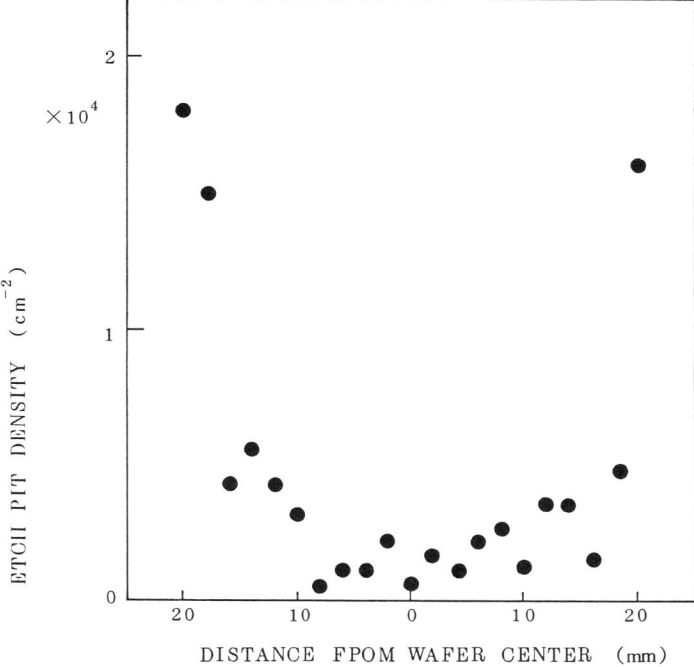

Fig. 7. Radial dislocation density profile across wafer cut from the shoulder of crystal 2.

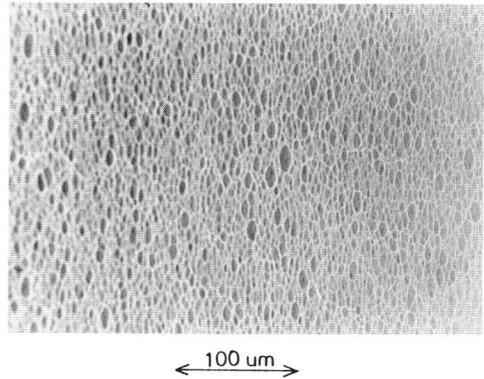

Fig. 8. Photomicrograph of KOH-etched wafer cut from crystal 2. This region is dislocation free.

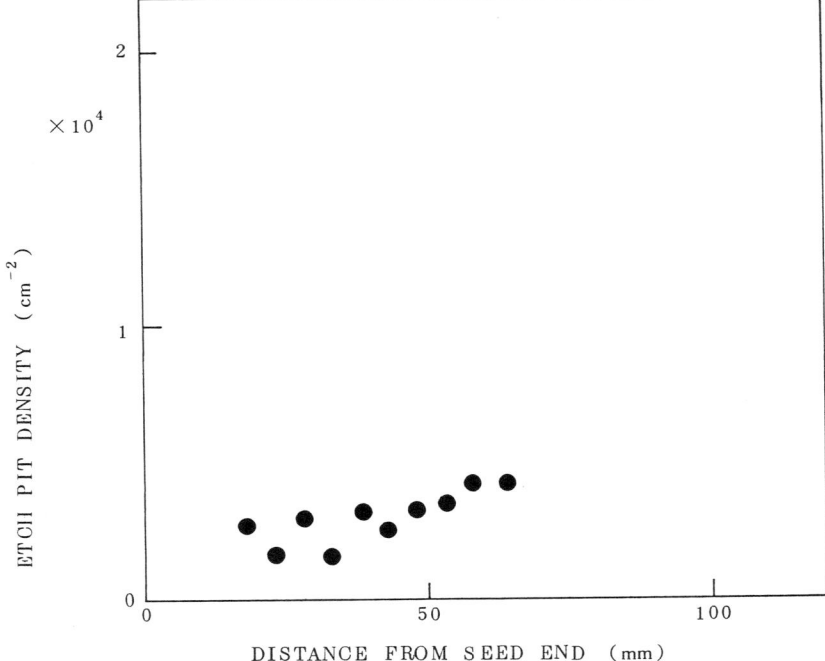

Fig. 9. The etch pit density distribution along the growth axis at an interval of 5 mm.

Ga droplets. As a result, generation of dislocations due to thermal stress was suppressed.

The results of evaluation of the electrical properties and impurity concentration are described below. Table 1 shows the results of Hall measurement, SIMS and LVM method of three crystals. Crystals 1 and 2 grown under an As atmosphere are both P type and the resistivity of each crystal is 1×10^4, 1×10^6 Ω-cm, respectively. On the other hand, crystal 3 grown by LEC method is N type and the resistivity is 1×10^7 Ω-cm. The differences of these electrical properties can be explained by the differences of impurity concentration as described in the following.

C concentration, an acceptor impurity, was evaluated by LVM method to be 7.6×10^{16}, 1.5×10^{16} and 0.8×10^{16} cm^{-3} in crystals 1 and 2 and an LEC crystal, respectively. Concentration of other acceptor impurities (Cr, Mn, Fe) were as low as $\sim 10^{15}$ cm^{-3} and no remarkable difference among these crystals was found. The concentration of donor impurities (Si, S) was lower than the C concentration. Consequently, it is reasonable to suppose that the electrical properties of these crystals are controlled by the C concentration and the deep

Table 1. Electrical propeties and impurity analyses of GaAs grown under various pulling conditions.

Sample №	Pulling Condition		RESISTIVITY		SIMS ANALYSIS (atms cm^{-3})							*1
	Atmosphere	B$_2$O$_3$	ρ (Ω-cm)	TYPE	B	Si	S	Cr	Mn	Fe	O	O
1	As	NO	1E4	P	1E15	1E16	3E15	1E15	7E14	5E15	1E16	7.6E16
2	As	YES	1E6	P	4E17	8E13	9E13	3E14	8E14	3E15	3E15	1.5E16
3	Ar	YES	1E7	N	4E16	1E14	6E14	1E15	2E15	1E16	3E15	0.8E16

*1 LVM method (atms cm^{-3})

donor El_2 concentration,[10] the resistivity is supposed to increase with decreasing C concentration in P type crystals.

It should be noted that the Si concentration of crystals 2 and 3 was much lower than that of crystal 1. This difference of the Si concentration can be explained by the gettering effect by B_2O_3 on the GaAs melt.[11]

The sources of C impurity seem to be carbon parts used in the quartz chamber. Therefore the another material should be used for the reduction of the C concentration.

Summary

A new apparatus was developed to grow GaAs crystals by the Czochralski method under a controlled As vapor pressure. The dislocation density in undoped (100) $2''^\phi$ crystals grown by this apparatus were as low as 2×10^3 cm^{-2}. Such a pronounced reduction of the dislocation density seemed to be associated with the improved melt composition resulting from the As vapor pressure control, the reduced thermal stress due to the low temperature gradient just above the melt and the diameter control without rapid change of the crystal diameter.

In addition, it was shown to be possible to grow the semi-insulating GaAs by the As vapor pressure controlling Czochralski method through the reduction of the C concentration.

REFERENCES

1) P. Petroff and R. L. Hartman: Appl. Phys. Lett. **23** (1973) 469.
2) Y. Nanishi, S. Ishida, T. Honda, H. Yamazaki, and S. Miyazawa: Jpn. J. Appl. Phys. **21** (1982) L335.
3) J. Nishizawa, Y. Okuno, and H. Tadano: J. Cryst. Growth **31** (1975) 215; J. Nishizawa and Y. Okuno, *Proc. Second Int. Sch. Semiconductor Optoelectron.* (Cetniewo, Poland, 1978); J. Nishizawa, N. Toyama, and Y. Oyama, *Proc. Third Int. Sch. Semiconductor Optoelectron.* (Cetniewo, Poland, 1981).
4) J. Van den Boomgaard and K. Schol: Philips Res. Rep. **12** (1957) 127.
5) A. S. Jordan, R. Caruso, and A. R. Von Neida: Bell Syst. Tech. J. **59** (1980) 593.
6) R. T. Chen and D. E. Holmes: J. Cryst. Growth **61** (1983) 111.
7) A. Steinmann and U. Zimmerli: J. Phys. Chem. Solids Suppl. **1** (1967) 81.
8) P. C. Leung and W. P. Allred: J. Cryst. Growth **19** (1973) 356.
9) J. Friedel: *Dislocations* (Pergamon, Oxford, 1964) pp. 92, 106.
10) A. T. Hunter, H. Kimura, J. P. Baukus, H. V. Winston, and O. J. Marsh: Appl. Phys. Lett. **44** (1984) 74.
11) T. Udagawa and T. Nakanishi: *Proc. Int. Symp. Gallium Arsenide and Related Compounds* (1981) p. 19.

Defects and Properties of Semiconductors: Defect Engineering, edited by J. Chikawa, K. Sumino, and K. Wada, pp. 37–69.
© KTK Scientific Publishers, Tokyo, 1987.

DEEP LEVEL PHOTOLUMINESCENCE IN GaAs

Michio TAJIMA*

Electrotechnical Laboratory, 1-1-4 Umezono, Sakura-mura, Ibaraki 305, Japan

Abstract. Deep level photoluminescence (PL) of undoped semi-insulating (SI) GaAs crystals grown by the liquid encapsulated Czochralski (LEC) method has been investigated in detail under various excitation conditions. The conventional above band-gap excitation (AGE) source induces the two typical PL bands; the 0.6-eV band with a peak ranging from 0.62 to 0.68 eV and with a half-width of 0.12–0.15 eV and the 0.8-eV band with a peak at 0.7–0.8 eV and with a half-width of about 0.3 eV. A comparison between the PL data and etch-pit observation suggests the correlation of the two bands with the dislocation and the melt composition. Although many researchers have reported the relationship between the 0.6-eV band and the main deep donor EL2, the definite identification of the band has not yet been made. We consider the variety in the 0.6-eV band is due to the fact that the origin of the EL2 level is not unique.

The analysis of PL under the below bang-gap excitation (BGE) is quite useful to solve the complex problem in the origin of the deep level PL. As for the BGE source, we use the 1.06-μm and 1.32-μm lines of a yttrium aluminum garnet laser. Both of the lines excite the deep level PL and the former line induces the PL fatigue (photoquenching) effect. The advantage of the 1.32-μm excitation is to excite the deep level PL band selectively without inducing the PL fatigue effect. As a result, the 0.67-eV band with a half-width of 0.12 eV has been observed commonly for undoped SI LEC crystals for the first time, regardless of the variety of the PL under the AGE. The 0.67-eV band is concluded to be associated with the same EL2 level as the one studied by the other techniques, such as the deep level transient spectroscopy (DLTS), the optical absorption spectroscopy, etc. The PL recovery effect has been observed for the first time for the fatigued 0.6-eV band under both the AGE and BGE conditions. This effect is explained by the transition from the metastable state to the normal state of the EL2 level. The fatigue and recovery rates are proposed to identify the deep levels responsible for the various 0.6-eV band. The PL spectroscopy under the simultaneous excitation of AGE and BGE reveals nonradiative deep centers in SI LEC crystals; these centers are never detected in boat-grown crystals.

We believe the present results will contribute greatly to a fundamental understanding of the complicated properties of the deep levels.

*Present Address: Optoelectronics Joint Research Laboratory, 1333 Kamikodanaka, Nakahara-ku, Kawasaki-shi 211, Japan.

1. Introduction

Deep levels in semiconductors are regarded as nuisance on one hand, since they have detrimental effects on device properties. On the other hand, deep levels act effectively as compensation centers for residual shallow impurities in semi-insulating (SI) substrate materials.[1,2] From either point of view the precise characterization of deep levels is eagerly required. Especially the assessment of midgap levels, the so-called EL2 levels, in undoped liquid encapsulated Czochralski (LEC) GaAs crystals has attracted a lot of attention because the SI property is essential to the development of high speed GaAs integrated circuits.[3] In spite of extensive works on the electrical and optical properties of the EL2 levels, the origin of the EL2 levels has not yet been clarified.[4] The characterization of the EL2 level in undoped SI LEC GaAs crystals has been performed mainly by the optical absorption[5-8] and photoluminescence (PL),[9-30] since conventional deep level transient spectroscopy (DLTS) is not applicable to SI materials.

In general, the PL technique has the advantage of capability of analyzing impurities and defects nondestructively with a high spatial resolution.[31,32] The sensitivity for various kinds of impurities and defects has been reported to be very high. However, the PL signals associated with the deep levels usually appear as featureless broad bands, which makes it difficult to perform the advanced analysis. Inapplicability of the PL technique to the assessment of nonradiative centers is also a disadvantage, since the recombination processes via the deep levels are often nonradiative.

In order to get more information on the deep levels we have measured and analyzed the PL under the below band-gap excitation (BGE) as well as under the above band-gap excitation (AGE). We have used the 1.06-μm line and the 1.32-μm line of the yttrium aluminum garnet (YAG) laser as the BGE source and the 0.5145-μm line of an Ar laser as the AGE source. What we have done are (1) comparison of PL under various excitation, (2) analysis of time dependent PL signal under alternative excitation of two kinds of excitation sources, and (3) analysis of PL signal under twofold excitations.

In this paper first we will briefly survey the history of deep level PL in GaAs. Then we will explain the optical transition associated with the deep levels using the configuration coordinate model in Sec. 3. Next we will show that the two PL bands, one with a peak at about 0.65 eV and the other with that at about 0.8 eV, dominate the deep level PL under the AGE condition in Sec. 4. A relationship between the intensities of these PL bands and the dislocation density will be discussed. Section 5 deals with the deep level PL under the BGE condition. We will show that a Gaussian band with a peak at 0.67 eV appears commonly in undoped SI LEC GaAs crystals and that the 0.67-eV band is correlated with the main deep donor EL2. In Sec. 6 we will describe the fatigue and recovery effects peculiar to the deep level PL associated with the EL2 level. The analysis of these effects gives us a prominent clue to clarify the metastable state of the EL2 level. Finally we will show the new technique called twofold excitation modulated photoluminescence (TEMPL), which makes it possible to analyze the nonradiative deep centers. We will demonstrate the crucial difference in the nonradiative deep centers between the LEC crystals and the boat-grown crystals. We believe that the present results will contribute greatly to a fundamental understanding of the complicated properties of the deep levels.

2. History of Deep Level PL in GaAs

A typical PL spectrum of undoped SI LEC GaAs at 4.2 K is shown in Fig. 1. The spectrum consists of three emission bands with a peak at 1.49, 0.80 and 0.65 eV. These three bands are commonly observed in undoped and lightly Cr-doped samples. The 1.49-eV band is due to the free-to-bound and donor-acceptor pair transition involving carbon acceptors.[33] The origins of the 0.80- and 0.65-eV bands are the main concern of the present work. The former band appears as a structureless broad band with a half-width of about 0.3 eV and with a peak around 0.8 eV. The latter band appears also as a featureless broad band with a peak ranging from 0.62 eV to 0.68 eV and with a half-width from 0.12 to 0.15 eV. The peak positions and the half-widths of the two bands depend on samples. Hereafter we call generally the former and the latter bands "the 0.8-eV band" and "the 0.6-eV band", respectively. When we discuss a particular band, we specify the peak position down to the second decimal place, e.g., "the 0.63-eV band." Occasionally, both of the 0.6- and 0.8-eV

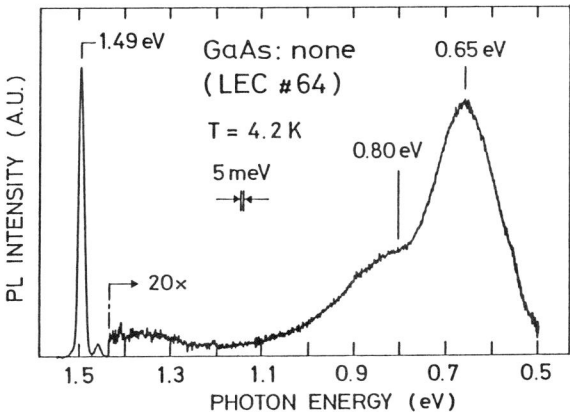

Fig. 1. PL spectrum at 4.2 K from undoped SI LEC GaAs crystal (sample #64). (Ref. 14)

bands appear with comparable intensities, which causes a very broad band with an asymmetric spectral shape as shown later (e.g., #2 in Fig. 6(a)). The two bands show a striking difference as we will explain in detail in this article.

The investigations on the 0.6- and 0.8-eV bands are summarized historically in Table 1. The 0.6-eV band was first reported by Turner et al.[9] At first the correlation between the 0.6-eV band and oxygen was investigated because oxygen was believed to act as a deep donor and to be present most commonly in GaAs crystals.[10,11] Then, the progress in the DLTS measurements indicated the presence of the so-called EL2 level.[34] The EL2 level attracted a lot of attention, since it acts as a compensation center for a residual shallow acceptor, through which SI property is realized.[1,2] The earlier studies often mixed the oxygen donor up with the EL2 donor, since there was no clear distinction between the oxygen level and the EL2 level until Huber et al.[35] claimed that the EL2 level is not related to oxygen in 1979. Then, many experimental results supported the idea that the EL2 level was ascribed to the As_{Ga} antisite.[4,36] In this context, many researchers tried to see the correlation among the 0.6-eV band, the EL2 level, and the As_{Ga} antisite defect.[13,14,16,17,22,27] For this purpose, the comparisons of their distributions in a crystal and their dependence on the melt composition were studied.

The characteristic photoquenching effect was observed for the EL2 level in photocapacitance[38,39] and optical absorption.[6,40,41] This effect is believed to occur as a result of the optical transition from the normal state to the metastable state of the EL2 defect. Therefore, the analysis of this effect gives us a prominent clue to determine the microscopic origin of the defect. The effect appears also in the 0.6-eV band,[15,19,23-25] which we call "PL fatigue effect." The PL fatigue effect is more complicated than the photoquenching effect in the photocapacitance and optical absorption, since the incident light plays two roles, i.e., the source for the above optical transition and the PL excitation source. Many controversial results have been reported as to the PL fatigue effect, reason for which will be described in Sec. 6.

The importance of the other deep level PL, the 0.8-eV band, was pointed out recently. The recent undoped SI LEC crystals often show the 0.8-eV band preferentially compared with the 0.6-eV band. As we will discuss later, the correlation between the 0.8-eV band and the EL2 level is considered to be negative. However, it is interesting to note that the appearance of the 0.8-eV band depends on the melt stoichiometry.[13,17,22]

At present the origins of the deep levels responsible for the 0.6- and 0.8-eV bands have not yet been settled. We believe, however, that the present work will contribute greatly to explain the contradictory results reported so far and to clarify the optical properties of the deep levels.

3. Optical Transition in EL2 Level

Optical transitions via the deep levels are well explained by the configuration-coordinate (CC) model.[42] Figure 2 shows the reported CC diagram for the EL2 level in GaAs.[39] For simplicity, we assume that there is a single EL2 level and that the radiative transition occurs between the EL2 level and the valence band. We consider that this transition is responsible for the 0.6-eV band.

For a conventional PL analysis, we use the light with a photon energy larger than the band-gap energy as an excitation source (AGE; above band-gap excitation). Electrons excited by the f process in Fig. 2(a) recombine directly with hole (g process; band-edge luminescence) or are captured by the EL2 level and then recombine with hole (h process; deep level PL). It should be noted that the deep levels are generally accompanied with large lattice relaxation. That is, the excited electrons displace the lattice through the strong electron-lattice interaction at the deep levels so that the total energy takes the minimum value. Therefore, the radiative recombination energy, E_h, is smaller than the thermal energy, E_0, which corresponds to the energy difference between the electrical energy level for the deep level and the valence band ($E_G - E_D$). This is called Franck-Condon shift. An emission spectrum appears as a featureless broad band with a peak at about E_h, and the half-width of the band becomes larger with increasing the electron-lattice interaction.

If the photon energy of the excitation source becomes smaller than E_G, the band edge emission (g) is no more observable, because the f process cannot occur. However, the deep level PL, h, can be induced by the optical absorption process, a, as shown in Fig. 2(b) and (c). This is called below band-gap excitation (BGE) PL. The electron-lattice interaction is also reflected in this process, which results in $E_a > E_0$ (Franck-Condon shift). In principle, the absorption spectrum (process a) and the emission spectrum (process h) show a mirror symmetry with respect to E_0.

The characteristic photoquenching effect is illustrated in b process in Fig. 2(b). This is explained by the optical transition from the normal state (EL2^0) to the metastable state (EL2*) with a very large lattice relaxation under the persistent illumination of light with a photon energy of $1.0 \leq h\nu \leq 1.4$ eV.[43] As a result of this transition, the PL intensity of the h process decreases gradually (the PL fatigue effect). Figure 2(c) shows the optical transition under the excitation of light with a photon energy of $0.8 \leq h\nu \leq 1.0$ eV. Under this excitation condition the a process is possible while the b process is impossible, which means that the deep level PL (the h process) is affected neither by the band-edge PL as in the case of Fig. 2(a) nor by the photoquenching effect as in the case of Fig. 2(b). Therefore, this excitation condition may be the most suitable condition for the detailed analysis of the deep level PL as shown in Sec. 5.

Table 1. History of Deep Level PL in GaAs.

Year	Name	Ref.	Conclusion	Basis
1963	Turner	9	0.65-eV: oxygen	Oxygen-doping
1978	Deveaud	10	0.65-eV: oxygen	Oxygen-implantation
1979	Yu	11	0.62-eV: oxygen	Oxygen-doping
1981	Mircea-Roussel	12	0.645-eV: EL2	Spectral shape analysis
1981	Yu	13	0.68-eV: AsGa 0.77-eV: GaAs	Melt composition
1982	Tajima	14	0.65, 0.68-eV: EL2 0.80-eV: microdefect	W-shaped radial profile M-shaped radial profile
1982	Leyral	15	0.65-eV: EL2	Fatigue effect
1982	Windscheif	16	0.8-eV: AsGa	Annealing property
1982	Tajima	17	0.8-eV: As-rich defect	Correlation with etch-pit
1982	Shanabrook	18	0.63-eV: nearest DA 0.68-eV: Dh	PL excitation spectroscopy
1982	Shanabrook	19	0.63-eV: EL2 0.68-eV: not EL2	Fatigue effect No fatigue effect
1982	Yu	20	0.63-eV: Oxygen 0.68-eV: EL2	Temperature dependence Frank-Condon shift

Year	Name	Ref.	Conclusion	Basis
1982	Yu	37	0.68-eV: EL2	Spectral shape analysis
1983	Wolford	21	0.64-eV: oxygen	Oxgen-doped $GaAs_{1-x}P_x$
1983	Kikuta	22	0.65-eV, 0.8-eV grouping	Crystal growth condition
1983	Kikuta	52	0.8-eV: As-rich condition	Melt composition
1984	Yu	23	0.68-eV: EL2	Fatigue effect
1984	Tajima	24	0.6-eV: not unique origin	Fatigue and recovery effects
1984	Pajet	25	0.64-eV: not EL2	Increasing effect (no fatigue)
1984	Tajima	26	Nonradiative deep center	PL under twofold excitation
1984	Kikuta	27	0.65, 0.8-eV: As_{Ga}	Fermi level position
1984	Samuelson	28	0.635-eV: not EL2	No fatigue effect
1985	Tajima	29	CC model for 0.6-eV PL	Recovery effect
1985	Tajima	30	0.67-eV: EL2	PL under BGE

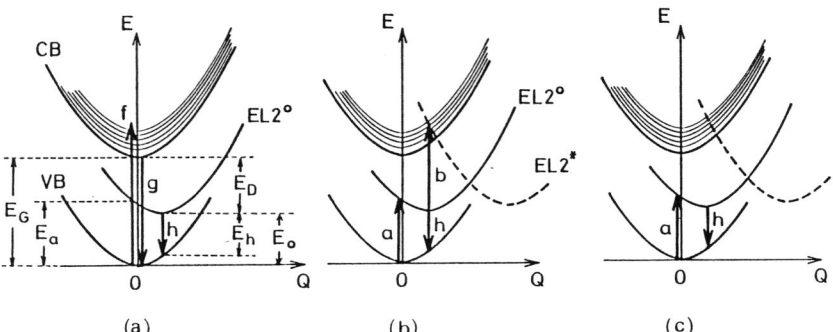

Fig. 2. Configuration coordinate diagram for EL2 level in GaAs. Possible optical transitions under various excitation conditions are shown by arrows. (a) $1.52\,\text{eV} < h\nu_{ex}$, (b) $1.0 \leq h\nu_{ex} \leq 1.4\,\text{eV}$, (c) $0.8 \leq h\nu_{ex} \leq 1.0\,\text{eV}$.

4. Deep Level PL under AGE

Figure 1 shows the typical PL spectrum of undoped SI LEC GaAs at 4.2 K measured under the AGE condition. The PL intensities of the three emission bands at 1.49, 0.80 and 0.65 eV vary along a wafer diameter as shown in Fig. 3(c). In order to see the correlation of PL intensities with the dislocation we performed the etch-pit observations using the AB solution and the molten KOH solution. Details of the etching/optical microscopy was reported previously.[44] Figure 3(a) shows the schematic illustration of the density profile for grooves and ridges revealed by the AB solution, and Fig. 3(b) represents the etch-pit density (EPD) profile obtained by the molten KOH etching. It should be noted that both grooves and ridges originate in dislocations and that ridges are accompanied with small pits which were identified as As precipitates.[45]

The 1.49- and 0.6-eV bands show the W-shaped intensity profile across a wafer diameter corresponding to the dislocation density, while the 0.8-eV band shows the M-shaped intensity profile. The origin of the 0.6-eV band was speculated to be due to the EL2 level from the following three reasons: The first is the positive correlation between the PL intensity and the dislocation density on a wafer. This is based on the fact that the density profile for the EL2 level coincides with the dislocation density in most cases, although this is not necessarily true at the tail end of a crystal.[7] Secondly the thermal energy obtained from the band shape analysis using the configuration coordinate model lies in the same energy region as the EL2 level (E_c-0.75 eV). Thirdly the PL fatigue effect occurs in the 0.6-eV band.

However, the recent studies have shown that the accurate peak position of the 0.6-eV band is ranging from 0.62 to 0.68 eV and the half-width is

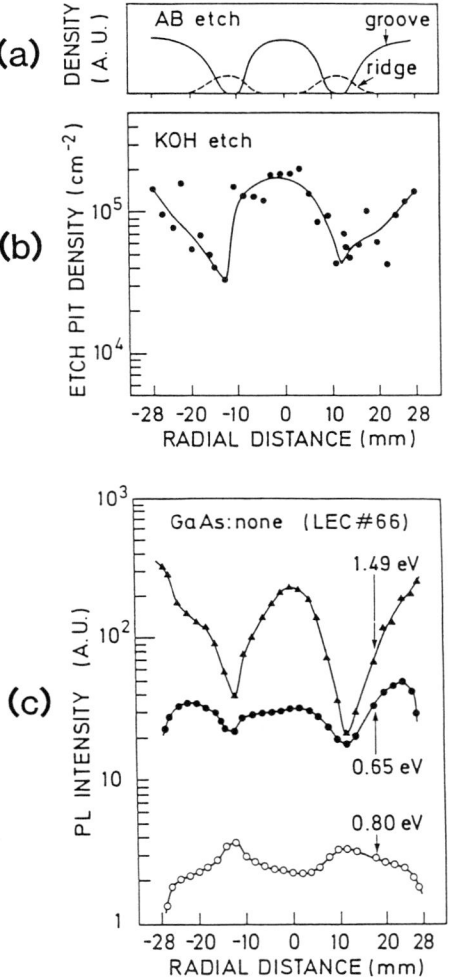

Fig. 3. Variation of etch-pit and PL intensity along wafer diameter for the undoped LEC GaAs wafer sliced from the tail side. (a) Schematic illustration of densities of ridges and grooves revealed by AB etching. (b) EPD revealed by molten KOH etching. (c) PL intensity profile. (Ref. 17).

distributed from 0.12 to 0.15 eV depending on a sample. Yu and Walters explained the above variation by the overlapping of the two bands; one with a peak at 0.63 eV which is related to oxygen and the other with that at 0.68 eV which is associated with the EL2 level. However, this idea is not sufficient to explain our experimental results. Figure 4 shows the PL spectra from three different SI LEC crystals at 4.2 K under the AGE condition. Although the peak positions of the three bands are distributed from 0.63 to 0.68 eV, their half-widths do not vary appreciably (about 0.15 eV for all the case). Furthermore, the magnitude of the PL fatigue effect differs from one sample to another.[29] These results suggest that the origin of 0.6-eV band is not unique, which will be discussed further in Sec. 5 and Sec. 6.

As for the 0.8-eV band, we believe that the EL2 level is not involved in the transition, because the above-mentioned three factors supporting the correlation with the EL2 level are all negative. Namely, the observed intensity distribution of the 0.8-eV band across a wafer diameter is contrary to the expected EL2 concentration. The thermal energy expected for the 0.8-eV band is about 1.0 eV, which is much larger than that for the EL2 level. The PL fatigue effect was never detected for the 0.8-eV band.

In the following part, we will discuss the origin of the 0.8-eV band. It has been pointed out by many workers[46-48] that in various semiconductors some microdefects, such as minute precipitates, vacancy complexes, dislocation loops, etc., are generated in the region where the dislocation density is decreased. This has been explained by assuming that the dislocations act as sinks for the microdefects. Such microdefects are also considered to appear in SI GaAs crystals, and their density will show an inverse profile with respect to the dislocation-density profile. Therefore, if we assume that the microdefects act as constituents of the radiative recombination centers responsible for the 0.8-eV band, the above profile for the 0.8-eV band can be explained. We can give an instance of microdefects which act as radiative recombination centers in heat-treated Si crystals.[49] If we further assume that the microdefects act as killer centers for the 1.49-eV band, the intensity profile for the 1.49-eV band can also be explained. In fact, it has been reported that such microdefects act

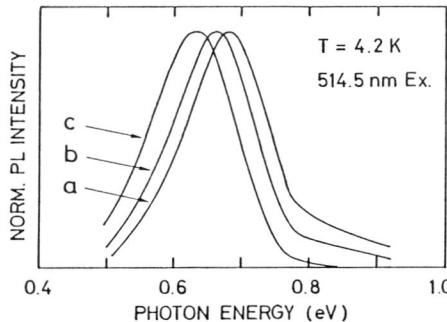

Fig. 4. Typical PL spectra of the 0.6-eV band for the three undoped SI LEC GaAs crystals at 4.2 K under the AGE condition. (Ref. 24).

as killer centers for efficient luminescence in GaAs crystals[50,62-66] and in an LEC-GaP crystal.[51]

We speculate that the microdefects responsible for the 0.8-eV band are formed preferentially in the As-rich condition during the crystal growth. This idea is based on the fact that the 0.8-eV band becomes strong in the region where the ridges are dominant, because the small pits on the ridges are reported to be As precipitates.[45] Our speculation contradicts to the conclusion given by Yu *et al.*[13] They concluded that the 0.77-eV band, which seems to be a member of the 0.8-eV band, is typical in Ga-rich grown crystals. A possible reason for this discrepancy is the variation of the PL spectra in the wafer as shown in Fig. 3(c), since they did not state the measurement positions in the wafers.

Kikuta *et al.* re-examined the relationship between the 0.8-eV band and the melt composition.[52] Their results show that the 0.8-eV band becomes strong with increasing the As fraction in the melt, which favors our speculation. These results lead us to suggest that the microdefects responsible for the 0.8-eV band is associated with the As_{Ga} antisite. This idea was supported by Windscheif *et al.*[16] on the basis of the coincidence of the annealing behavior between the 0.8-eV band and the EPR signal of the As_{Ga} antisite defect.

The typical relationships between the variations of PL intensities and the dislocation density profile on a wafer is summarized in Fig. 5. The other optical and electrical properties are also shown in the figure.[67] The dislocation density profile is W-shaped across a wafer diameter as shown in (a), which

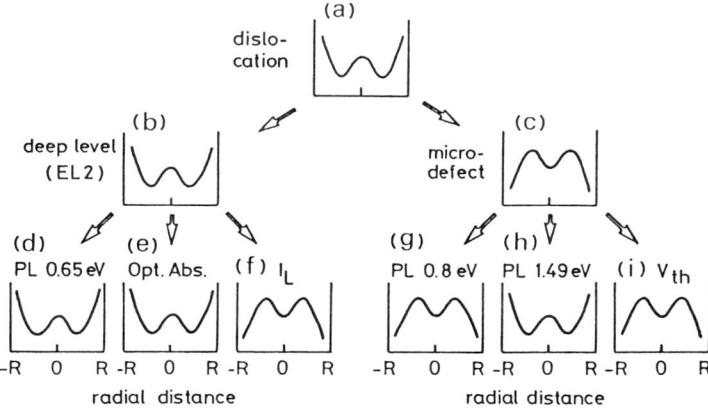

Fig. 5. Schematic illustration of correspondence of radial variation between dislocation density and various properties of deep levels.

implies that the dislocations are induced by thermal stress.[68] The 0.6-eV PL band and the 1-μm optical absorption are considered to be associated with the EL2 level. Both of the two distributions along a wafer diameter coincide with the dislocation density profile as shown in (d) and (c). This result leads us to suggest that the formation of the microdefects responsible for the EL2 level is enhanced by the thermal stress or by the dislocation motion. The measurement of leakage current I_L is widely used to check the inhomogeneity of SI wafers. The leakage current is reported to be inversely proportional to the deep trap density. The W-shaped density profile for the EL2 level explains the M-shaped I_L profile.[67]

On the other hand, the dislocations act as gettering sites for some microdefects. This explains the intensity profiles for the 1.49-eV band (g) and for the 0.8-eV band (h), as described in the preceding part. The microdefects are considered to affect detrimentally the activation of implanted impurities in fabricating field effect transistors (FET's). Therefore, the variation of the threshold voltage of FET, V_{th}, corresponds to the microdefect distribution, as shown in (i).[67]

5. Deep Level PL under BGE

As we discussed in the preceding section, two deep level PL bands appear in undoped SI LEC GaAs crystals under the AGE condition; the 0.6-eV band and the 0.8-eV band. The intensity distribution on a wafer for the former band is W shaped along a radial distance, while the latter band is M shaped. Since the intensity ratio between the 0.6-eV and 0.8-eV bands differs greatly depending on the samples, undoped SI LEC crystals are classified into two groups according to the following: either the 0.6-eV band or the 0.8-eV band dominates the PL spectrum, the 0.6-eV band dominant crystals and the 0.8-eV band dominant crystals.[13,22] The peak position of the 0.6-eV band varies from 0.62 to 0.68 eV depending on the samples, as we pointed out in Sec. 4.

In spite of the above-mentioned variety in the PL results, no remarkable differences have been detected in the results obtained by the optical absorption and resistivity analyses for undoped SI LEC crystals.[53] For example, both the 0.6-eV band dominant and the 0.8-eV band dominant crystals contain the EL2 level with a concentration of about 1×10^{16} cm^{-3} with the W-shaped density profile on a wafer. We consider that the above discrepancy between the PL and the other analyses originates from the fact that there are many species of deep levels in undoped LEC crystals and that only a few of them are commonly present in the crystals. In conventional PL analysis under the AGE, most levels in the band gap are photoexcited. As a result of competitions among various recombination processes, certain transitions with high radiative recombination probability dominate the PL spectrum. These transitions do not necessarily involve the deep levels commonly present in undoped LEC crystals. This explains the wide variety of deep level PL under the AGE.

From this viewpoint, we tried to characterize the deep levels which are commonly present in undoped SI LEC GaAs crystals but are not necessarily detectable by the conventional PL under the AGE. We have used the 1.32-μm line of a YAG laser as a BGE source. As we described in Sec. 3 this excitation has two great advantages. First, this light can excite only midgap levels, i.e., the photon energy of this light is too small to excite the levels responsible for the near-band-edge emission or the 0.8-eV band. As a result, the recombination process becomes much simpler than in the case of the AGE. Second, the PL fatigue effect does not occur for this photon energy.

The samples used for the present study are eight undoped (100) SI LEC GaAs wafers obtained from four different suppliers. Samples were immersed in liquid He and excited by the 514.5-nm line of an Ar laser, the 1.06-μm line or the 1.32-μm line of a YAG laser with an incident power of about 1 W·cm^{-2}. After the spectral response calibration, the spectra were plotted with the abscissa and the ordinate linearly proportional to the photon energy and the number of photons, respectively.

Three typical types of PL spectra measured under Ar laser excitation at 4.2 K are shown in Fig. 6(a). These spectra were taken at the center part of the three different wafers, and the intensities were normalized at their respective peak heights. In sample No. 1 the spectrum is dominated by an emission band with a peak at 0.74 eV and a full width at half-maximum (FWHM) of 280 meV (the so-called 0.8-eV band). The intensity distribution of this band is M shaped along a radial distance, as shown in Fig. 2. This is an inverse profile with respect to the etch-pit density (EPD) distribution, which coincides with the results described in Sec. 4. The 0.6-eV band is so weak that it is buried in the strong 0.74-eV band except for the edge part of the wafer. In sample No. 3 an emission band with a peak at 0.66 eV and a FWHM of 150 meV (the so-called 0.6-eV band) dominates the spectrum. This band exhibits a W-shaped intensity profile across a wafer diameter corresponding to the EPD distribution. This is also consistent with the previous results. In sample No. 2 the spectrum is considered to consist of both of the above two bands, which makes an asymmetric emission band with a long tail in the higher energy side.

When these samples are excited with the 1.32-μm line of a YAG laser, the respective samples show the spectra with the same shape as shown in Fig. 6(b). The traces of the three normalized spectra coincide with each other within our experimental errors. The peak position and the FWHM are 0.67 eV and 120 meV, respectively. This band is commonly observed for most samples (six out of eight samples) under this excitation condition. This band is apparently different from the 0.68-eV band measured under the AGE by Yu.[20,23,37] The difference in the peak position of 0.01 eV is much larger than our experimental error. One of our samples shows a Gaussian band with a peak at 0.68 eV and a

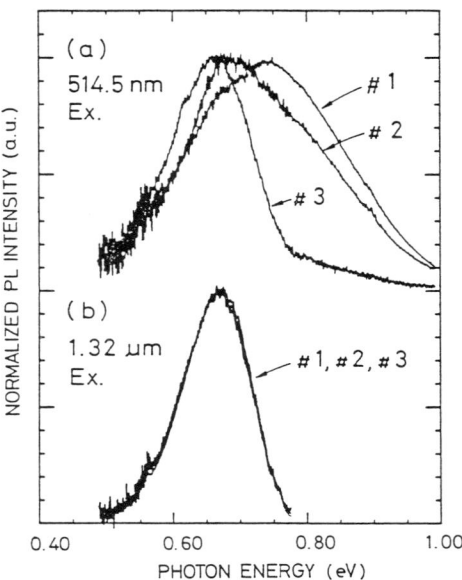

Fig. 6. (a) Typical PL spectra of undoped SI LEC GaAs crystals (samples No. 1, 2, and 3) measured under 514.5-nm excitation at 4.2 K. (b) PL spectra of the same samples measured under 1.32-μm excitation at 4.2 K. All the spectra are normalized at their respective peak heights. (Ref. 30).

FWHM of 130 meV under the AGE, which seems to be identical with the 0.68-eV band reported by Yu. This sample also exhibits the same 0.67-eV band under the 1.32-μm light excitation.

The radial intensity distribution of the 0.67-eV band of sample No. 1 excited with the 1.32-μm light is shown in Fig. 7. The intensity distribution is W shaped in contrast with the M-shaped profile for the 0.74-eV band measured under the AGE. The W-shaped intensity profile for the 0.67-eV band has a positive correlation with the EPD distribution. As for the other samples, the intensity profile for the 0.67-eV band is always W shaped along a radial distance corresponding to the EPD distribution. The W-shaped EL2 density profile has been equally observed for both the 0.6-eV band dominant and the 0.8-eV band dominant crystals by the optical absorption analysis. This supports the idea that the 0.67-eV band is associated with the EL2 level.

In order to investigate the fatigue effect of the 0.67-eV band, we have taken the following procedure: (i) irradiate the sample with the 1.32-μm light and measure the initial intensity, (ii) irradiate the sample with the 1.06-μm light for 30 s, (iii) irradiate the sample with the 1.32-μm light and measure the intensity. Procedure (ii) induces the transition from the normal state to the

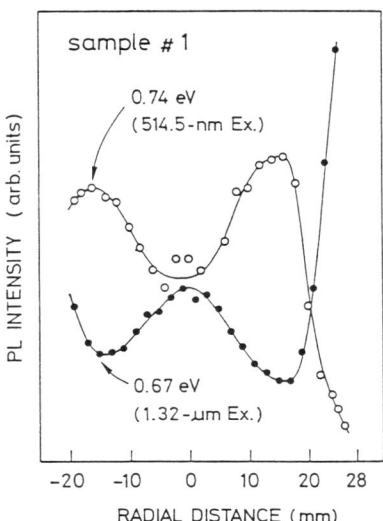

Fig. 7. PL intensity profiles along a radial distance for 0.74-eV band (○) measured under 514.5-nm excitation and 0.67-eV band (●) under 1.32-μm excitation in undoped SI LEC GaAs wafer (sample No. 1). (Ref. 30).

metastable state of the EL2 level.[9] What was observed by the procedure is that the PL intensity decreases to less than one-tenth of the initial intensity. The intensity is recovered to the initial intensity if we raise the sample temperature higher than 150 K for about 5 min. This corresponds to the thermal de-excitation process from the metastable state to the normal state.[9] This is the manifestation of the characteristic fatigue effect peculiar to the EL2 level.

A deep level PL band with a Gaussian shape is well analyzed with the configuration coordinate model.[42] A Franck-Condon shift, a difference between the thermal energy and the peak energy of the Gaussian band, can be obtained from a FWHM of the Gaussian band at 0 K and the coupled phonon energy.[12,13] If we assume that the coupled phonon energy is 22 meV according to previous works on the spectral shape analyses for the 0.645-eV band[12] and the 0.68-eV band[13,37] the Franck-Condon shift is determined to be 0.12 eV. If we further assume that the 0.67-eV band is due to a transition between the valence band and a deep donor, the donor level is estimated to be located about 0.73 eV below the conduction band. This value is close to the thermal ionization energy of the EL2 level.[54]

Although most samples (six out of eight samples) exhibit the same 0.67-eV band under the 1.32-μm excitation, the rest of the samples show a different emission band with a peak at 0.655 eV and a FWHM of 140 meV. The intensity profile of the 0.655-eV band along a radial distance is W shaped.

The fatigue effect is observed in the same way as the case for the 0.67-eV band. The spectral shape indicates that the donor level is located about 0.69 eV below the conduction band. These results show that the 0.655-eV band is also related to the EL2 level. This leads us to suggest that at least two kinds of the EL2 levels are present in undoped SI LEC GaAs crystals.

In conclusion, we have observed a Gaussian PL band with a peak at 0.67 eV and a FWHM of 120 meV under excitation with the 1.32-μm light. This band appears commonly in undoped SI LEC GaAs crystals regardless of the groups of the crystals classified according to their dominant PL band measured under the AGE. The spectral shape of the 0.67-eV band, the intensity distribution on a wafer, and the PL fatigue effect indicate that the radiative transition responsible for the 0.67-eV band involves the deep donor EL2 commonly present in undoped LEC GaAs crystals.

6. Fatigue and Recovery Effects in Deep Level PL

The PL fatigue effect was first reported by Leyral et al. for the 0.65-eV band.[15] As described in Sec. 4, the peak energy of the 0.6-eV band carefully examined varies in the range between 0.62 and 0.68 eV depending on the sample. Further studies on the fatigue effect on the 0.62–0.68-eV band have shown contradictory results. Shanabrook et al.[19] observed the fatigue effect for the 0.63-eV band but not for the 0.68-eV band. On the contrary Samuelson et al. did not observe the fatigue effect for the 0.63-eV band[28] and Yu observed that for the 0.68-eV band.[23]

Here we have to keep in mind that the deep level PL can be induced under the BGE condition as well as the AGE condition. The PL fatigue effect was reported to occur under the light irradiation with a photon energy between 1.0 and 1.4 eV,[43] as shown in the solid line in Fig. 8. Leyral et al.[15] and Shanabrook et al.[19] measured the PL fatigue effect under the 1.06-μm light irradiation. In this case, irradiated light acts as the excitation source for the deep level PL on one band and as that for the optical transition responsible for the fatigue effect on the other band. In contrast with this, Yu[23] and Samuelson et al.[28] measured the PL fatigue effect by comparig the PL under the AGE condition before and after the light irradiation which causes the optical transition responsible for the fatigue effect. The origin of the PL under the AGE condition and that under the BGE condition are not necessarily the same. The above-mentioned difference in the experimental procedure is considered to account for the contradictory results.

In this work we reinvestigate carefully the fatigue effect of the 0.6-eV band under both AGE and BGE conditions. In the course of the study, we have found for the first time the characteristic recovery effect, which occurs for the "fatigued" PL bands by continuous irradiation of AGE or BGE light. The mechanism of the recovery effect is discussed using a configuration

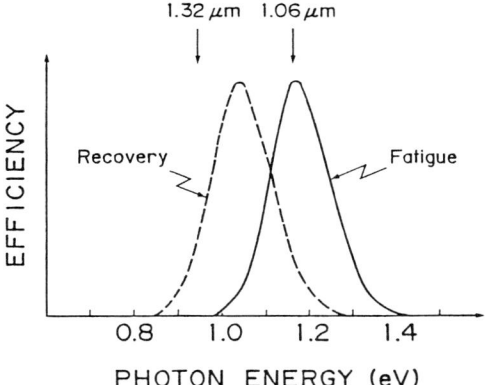

Fig. 8. Spectral distribution of efficiency for PL fatigue effect (EL2^0→EL2*, solid line, Ref. 43). Broken line represents expected transition probability for EL2*→EL2^0.

coordinate (CC) diagram. We show that the recovery effect occurs as a result of the existence of the metastable state with large lattice relaxation. We propose fatigue and recovery rates which can identify the deep levels responsible for "various" 0.6-eV bands. We will discuss the microscopic structure of microdefects responsible for the EL2 level.

The samples used for the present work were commercially available undoped (100) SI LEC GaAs wafers obtained from four different sources. We selected crystals in which the 0.6-eV band was dominant, in order to avoid the contribution of the 0.8-eV band. We primarily measured the center part of the wafer. The samples were cooled down to 4.2 K in the dark in order to prevent inadvertent irradiation of the light which might induce the fatigue effect. We initialized a sample which had experienced the fatigue effect by an anneal at 150 K for 10 min (initialization process). As excitation sources we used the 0.5145-μm (2.41-eV) line of an Ar laser and 1.06-μm (1.17-eV) and 1.32-μm (0.94-eV) lines of a YAG laser with a typical incident power of about 1 W·cm^{-2}.

In this article we will describe the typical result on one sample (sample #07). All of the above three excitation sources can induce the 0.6-eV band, as shown in Fig. 9. In order to specify the respective PL we use such symbol as PL$_{1\cdot32}$, where we indicate the wavelength of the excitation light in μm in subscript. The peak positions of the PL$_{0\cdot5145}$, PL$_{1\cdot06}$, and PL$_{1\cdot32}$ band are 0.66, 0.68, and 0.67 eV, and their half-widths are 0.14, 0.13, and 0.12 eV, respectively. All the spectra were taken after the initialization process, and the spectrum under the 1.06-μm light irradiation was measured after sufficient exposure since this light induces the PL fatigue effect. Hereafter we will discuss the three types of fatigue and recovery effects.

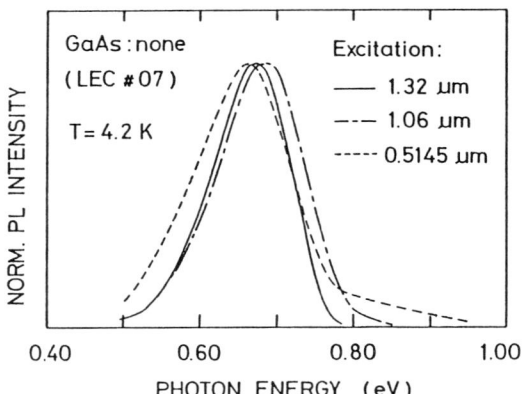

Fig. 9. PL spectra of an undoped SI LEC GaAs crystal at 4.2 K under 0.5145-μm, 1.06-μm and 1.32-μm light excitation. Peak intensities are normalized to unity. (Sample #07) (Ref. 29).

The first type of the fatigue and recovery effects was observed under the alternative excitation of 0.5145-μm and 1.06-μm light as shown schematically in Fig. 10. We denote T_i the time when the i-th excitation light is turned on, t_i the period of the i-th excitation light, and $I(t)$ the peak PL intensity at t.

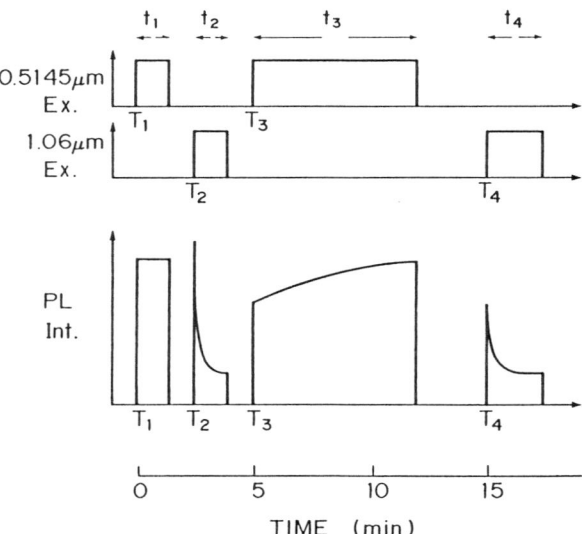

Fig. 10. PL fatigue and recovery effects induced by alternative excitation of the 0.5145-μm and 1.06-μm light. For the PL intensity under 0.5145-μm or 1.06-μm light excitation, the peak intensity of the 0.66-eV or 0.68-eV band shown in Fig. 9 is plotted. (Sample #07).

The 0.5145-μm light, turned on at $t=T_1$, excite the 0.66-eV $PL_{0.5145}$ band with a peak intensity of $I(T_1)$. This intensity decreases rapidly as a function of time during the 1.06-μm light irradiation ($T_2 \leq t \leq T_2+t_2$); this is the fatigue effect for the 0.68-eV $PL_{1.06}$ band. The decay profile can be roughly approximated by the exponential decay with a sample dependent time constant ranging from 1 to 5 s for the present excitation condition. In all cases, the intensity reaches a constant value within a few minutes. When the sample is irradiated again with the 0.5145-μm light at $t=T_3$, the rise of PL intensity, $I(T_3)$, is lower than the initial PL intensity, $I(T_1)$. This is the fatigue effect for the 0.66-eV $PL_{0.5145}$ band which was induced by the 1.06-μm light irradiation. The PL intensity is restored gradually towards the initial intensity by the continuous 0.5145-μm light irradiation; this is the recovery effect for the 0.66-eV $PL_{0.5145}$ band induced by the 0.5145-μm light. The temporal recovery is not purely exponential. The time required to restore the initial intensity varies from 20 min to several hours depending on the sample. The fatigue effect is again observed for the 0.68-eV $PL_{1.06}$ band after the recovery procedure ($T_4 \leq t \leq T_4+t_4$). The rise in the PL intensity, $I(T_4)$, reflects the degree of recovery effect for the 0.68-eV $PL_{1.06}$ band which was induced by the 0.5145-μm light irradiation. The $I(T_4)$ value approaches $I(T_2)$ with increasing time, t_3, of the pre-irradiation by the 0.5145-μm light. In most samples, the fatigue and recovery effects, induced by the 1.06-μm light and the 0.5145-μm light irradiations respectively, take place in both the $PL_{0.5145}$ and $PL_{1.06}$ bands.

The second type of the fatigue and recovery effects was observed under the alternative excitation of 1.32-μm and 1.06-μm light as shown schematically in Fig. 11; all the measured samples showed this effect. The 1.32-μm light, turned on at $t=T_1$, excites the 0.67-eV band with a peak intensity of $I(T_1)$. When the 1.06-μm light is switched on at $t=T_2$, the 0.68-eV band is induced with the intensity decreasing from the maximum value $I(T_2)$ to $I(T_2+t_2)$. This is the fatigue effect for the 0.68-eV $PL_{1.06}$ band. Then we irradiate the sample with the 1.32-μm light again at $t=T_3$. The rise in the PL intensity, $I(T_3)$, is less than 1/10 of $I(T_1)$, which represents the fatigue effect for the 0.67-eV $PL_{1.32}$ band. This PL intensity increases exponentially with a time constant of 5.2 min. About 20-min irradiation restores the PL intensity to the initial value. This is the recovery effect for the 0.67-eV $PL_{1.32}$ band. If we then turn on the 1.06-μm light at $t=T_4$, we observe the same fatigue effect as the previous one at $t=T_2$. This means the fatigued intensity $I(T_2+t_2)$ is restored to $I(T_4)$ by the 1.32-μm irradiation. This substantiates the recovery effect for the 0.68-eV $PL_{1.06}$ band.

The third type of the fatigue and recovery effects is shown schematically in Fig. 12; this type did not necessarily appear for all the samples. In this case we used only the 1.06-μm light as the excitation source and alternated the intensity. The intensity ratio between the strong and weak excitations is 1:0.1. The initial strong 1.06-μm light excites the 0.68-eV band accompanied by the

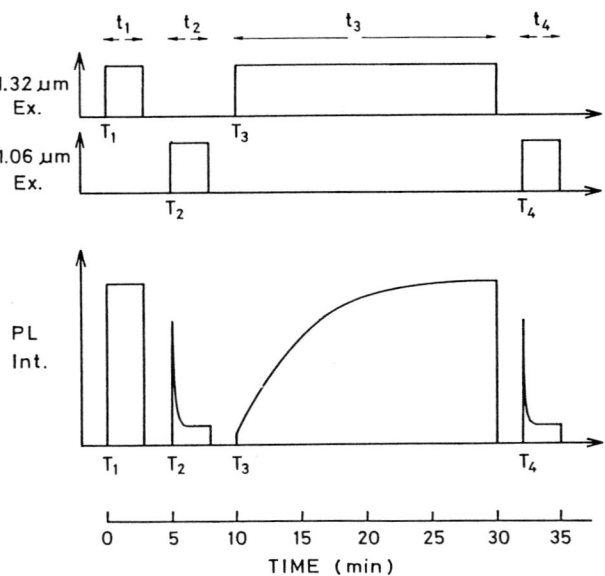

Fig. 11. PL fatigue and recovery effects induced by alternative excitation of the 1.06-μm and 1.32-μm light. For the PL intensity under 1.06-μm or 1.32-μm light excitation, the peak intensity of the 0.68-eV or 0.67-eV band shown in Fig. 9 is plotted. (Sample #07) (Ref. 29).

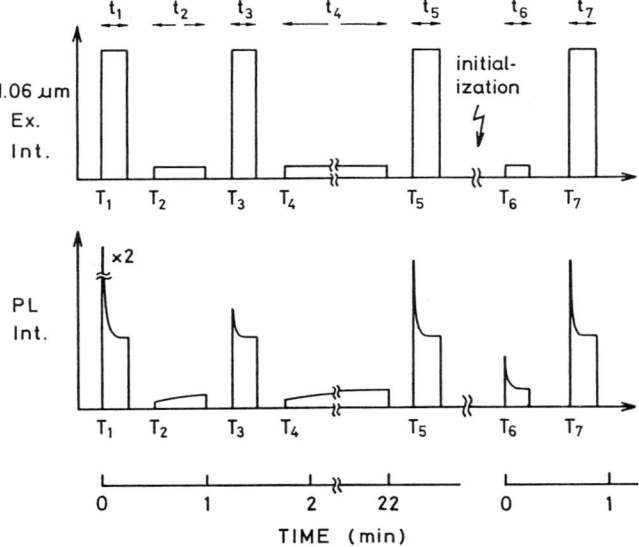

Fig. 12 PL fatigue and recovery effects of the 0.68-eV band induced by 1.06-μm light excitation, changing the intensity alternatively with the intensity ratio of 1:0.1. (Sample #07) (Ref. 29).

fatigue effect. If the weak excitation is turned on at T_2, the initial rise in the PL intensity is about 1/10 of $I(T_1+t_1)$ after which it increases gradually. It should be noted that the same excitation induces the fatigue effect after the initialization process as shown in $T_6 \leq t \leq T_6+t_6$. When we switch on the strong excitation again at $t=T_3$, the rise in PL intensity is higher than the fatigued intensity $I(T_1+t_1)$ but lower than the initial intensity $I(T_1)$. This means that a part of the PL intensity is restored by the weak excitation. This is the other newly observed recovery effect for the 0.68-eV $PL_{1.06}$ band. The degree of recovery becomes larger for a longer period of the weak excitation as shown in the figure. The degree also becomes larger for a smaller intensity ratio of the weak excitation to the strong excitation, although the period required for maximum recovery becomes longer.

In the following part we will discuss the mechanism of the recovery effect qualitatively using the CC model. The PL intensity at respective stages in Figs. 10-12 reflects the concentration of the $EL2^0$ state since the observed PL is considered to be due to the radiative transition between the $EL2^0$ state and the valence band. The fatigue effect has been well explained[38,39] by the optical transition for an electron from the normal state ($EL2^0$) to the metastable state ($EL2^*$) in a CC diagram, as shown by the arrow b in Fig. 13. This transition occurs for photo-excitation with energies between 1.0 and 1.4 eV, as shown in Fig. 8. The transition probability for the 1.06-μm light is close to the maximum value, while that for the 0.5145-μm light and that for the 1.32-μm light are very small.

Now we speculate the transition probability for the $EL2^*$ to $EL2^0$ (arrow d), which seems to be responsible for the recovery effect. Judging from the CC

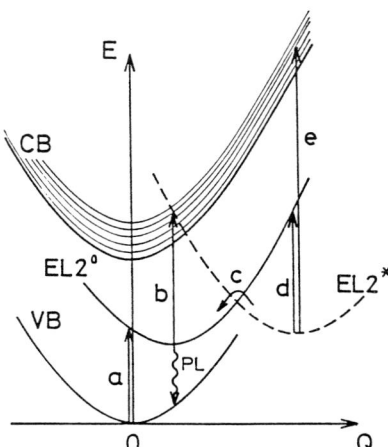

Fig. 13. Configuration coordinate diagram for the EL2 level in GaAs. $EL2^0$ and $EL2^*$ represent the normal and metastable states. Various transitions (arrows) are explained in the text. (Ref. 29).

diagram in Fig. 13, the EL2* to EL2⁰ transition is expected to be induced by the photo-excitation analogous to the EL2⁰ to EL2* transition. Considering that the relative position of the CC curve for the EL2* state is slightly higher in the E-axis direction than that for the EL2⁰ state, the spectral distribution of the EL2* to EL2⁰ transition probability is speculated to shift slightly toward the lower photon energy side, as shown in the broken line in Fig. 8.

The second type recovery effect shown in Fig. 11 supports this idea. The photon energy of the 1.32-μm light (0.94 eV) is not large enough to induce the EL2⁰ to EL2* transition but is sufficient for the EL2* to EL2⁰ transition. Since the EL2* states are created by the 1.06-μm light irradiation, the succeeding 1.32-μm light irradiation restores the EL2* states to the EL2⁰ states.

Figure 8 suggests the 1.06-μm light induces both the EL2⁰ to EL2* and the EL2* to EL2⁰ transition. The occurrence of the EL2* to EL2⁰ transition by the 1.06-μm light is clearly evidenced by the third type recovery effect shown in Fig. 12. The weak excitation is considered to enhance the relative rate of the EL2* to EL2⁰ transition with respect to the EL2⁰ to EL2* transition. Further investigations on the excitation intensity dependence of the transition rates of the a, b, d, and PL transitions in Fig. 13 will be necessary for quantitative understanding of the fatigue and recovery effects.

The EL2⁰ state will be converted entirely to the EL2* state under the 1.06-μm light irradiation, if the EL2* to EL2⁰ transition is not taken into account. This contradicts the experimental fact that the steady state PL intensity does not go to zero (e.g., $I(T_2+t_2)$ in Figs. 10 and 11). The steady state PL intensity can be explained alternatively by the presence of the overlapping PL bands which do not show the fatigue effect. However, this idea cannot account for the third type recovery effect.

The EL2* to EL2⁰ transition will not occur under the 0.5145-μm light irradiation, as shown in Fig. 8. However, the photon energy of this light seems to be large enough to realize the optical transition from EL2* to conduction band (arrow e in Fig. 11). We consider that this transition is responsible for the first type recovery effect.

The present recovery process is related neither to the thermal process (arrow c in Fig. 13) nor to the free carrier interaction (the Auger-like de-excitation[55]), which were reported so far to be responsible for the de-excitation process from the EL2* to the EL2⁰ state. The probability of the thermal de-excitation is negligible since samples are maintained at 4.2 K. The concentration of the excited free carriers is very low since the excitation sources are BGE in the second or the third type of the recovery effect. Correspondingly, near band-edge emission was below the detection limit. Furthermore, the recovery effect should be greatly enhanced by the above band-gap excitation if the free carrier interaction governs the process. Actually, however, no remarkable difference was observed between the first type recovery effect induced by AGE and the second and third type recovery effects induced by BGE.

The temporal dependence of the fatigue and recovery effects and the magnitudes of these effects differ from one sample to another. Here we introduce the following four rates for characterizing the fatigue and recovery effects of the first and the second type:

$$F_{\lambda_A} = (I(T_1) - I(T_3))/I(T_1),$$

$$F_{\lambda_B} = (I(T_2) - I(T_2 + t_2))/I(T_2),$$

$$R_{\lambda_A} = (I(T_3 + t_3) - I(T_3))/(I(T_1) - I(T_3)),$$

$$R_{\lambda_B} = (I(T_4) - I(T_4 + t_4))/(I(T_2) - I(T_2 + t_2)),$$

where $I(t)$ is the PL intensity at t as shown in Figs. 10 and 11. The fatigue and recovery rates are given by F_{λ_A} and R_{λ_A} for the PL band excited by the light with a wavelength of λ_A, which induces the recovery effect ($\lambda_A = 0.5145$ or 1.32 μm in the present experiment). The fatigue and recovery rates are given by F_{λ_B} and R_{λ_B} for the PL band excited by the light with a wavelength λ_B, which induces the fatigue effect ($\lambda_B = 1.06$ μm in the present experiment). These rates take a value between 0 and 1, where 0 and 1 correspond to no and complete fatigue (recovery), respectively. The recovery rate can be defined only for the samples that show the fatigue effect. In Table 2 we summarize these rates, where t_2 is 1 min, t_3 for the 0.5145-μm light irradiation is 12 min, and that for the 1.32-μm light irradiation is 20 min.

We can speculate on the origin of the EL2 level from the present results. A change in the spectral shape of the 0.6-eV band is explained by a variation of lattice relaxation in the normal state of the EL2 level, since the radiative transition for an electron is from the normal state to the valence band as shown in Fig. 13. On the other hand, a difference in the fatigue and recovery rates is interpreted a variety of lattice relaxations in the metastable state. The present data show that a wide variety of lattice relaxations takes place in both normal and metastable states. Recent studies have shown that the formation of the EL2 level is enhanced by increasing the As fraction in the melt,[36] that excess As atoms are present in undoped SI LEC GaAs crystals,[58] and that amorphous precipitates are detected by the transmission electron microscopy.[59] These facts favor the idea[56,57] that the microdefect responsible for the EL2 level is As atom aggregates, because a variety of lattice relaxation is explained by various possible configurations of As atoms in the lattice.

In conclusion, we have examined the characteristic fatigue and recovery effects of the 0.6-eV band associated with the EL2 level in undoped SI LEC GaAs crystals. The recovery effect is explained as a result of an optical transition from the metastable state to the normal state of the EL2 level and

Table 2. Fatigue and recovery rates for the 0.6-eV band.

Ex. Mode	0.5145 μm/1.06 μm						1.32 μm/1.06 μm					
PL Ex.	0.5145 μm			1.06 μm			1.32 μm			1.06 μm		
Sample	Peak energy (eV)	$F_{0.5145}$	$R_{0.5145}$	Peak energy (eV)	$F_{1.06}$	$R_{1.06}$	Peak energy (eV)	$F_{1.32}$	$R_{1.32}$	Peak energy (eV)	$F_{1.06}$	$R_{1.06}$
a	0.68	0	–	0.68	0.6	0.1	0.67	*	*	0.68	*	*
b	0.66	0.3	0.8	0.68	0.8	0.5	0.67	0.9	1.0	0.68	0.8	1.0
c	0.63	0.2	0.6	0.62	0.6	0.4	0.655	1.0	1.0	0.62	0.6	1.0

– undefined; *not measured.

that from the metastable state to the conduction band. A wide variety of the fatigue and recovery rates indicates that various lattice relaxations are induced by the microdefects responsible for the EL2 level. These results favor the idea that the EL2 level is due to As atom aggregates.

7. Characterization of Nonradiative Deep Levels

As shown in the preceding sections, the PL technique is very powerful for characterizing deep levels in GaAs. However, the technique has a serious disadvantage of inapplicability to nonradiative recombination centers. In order to overcome the problem we have developed a new technique called twofold excitation modulated photoluminescence (TEMPL) spectroscopy to study nonradiative recombination process as well as radiative recombination process in deep levels in semiconductors.[28]

The TEMPL spectroscopy is based on the analysis of PL variation induced by the below band-gap excitation (BGE) in addition to the above band-gap excitation (AGE) using modulation technique. In order to illustrate the excitation and recombination processes under twofold excitation, we take a simplified system in which we assume (i) only one deep donor level is present, (ii) recombination processes are a radiative transition between the conduction band and the valence band and that between the donor and the valence band, (iii) the photon energy of the BGE light is large enough to cause a transition from the donor to the conduction band and that from the valence band to the donor, and (iv) the probability for a two-step excitation process via the donor under the BGE is negligibly small.

A schematic picture of the excitation and recombination processes is shown in Fig. 14. The AGE generates excess electrons and holes (process a in Fig. 14). The electrons recombine with holes (b) or they are trapped by the

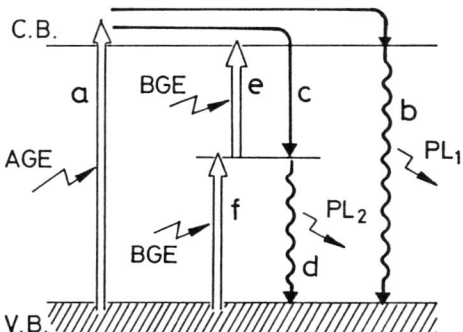

Fig. 14. Schematic illustration of excitation and recombination processes under twofold excitation. (Ref. 26).

donors (c) followed by recombination with holes (d). These recombination processes cause band-to-band emission, PL_1, and donor-to-band emission, PL_2, respectively. When the BGE is additionally turned on, electrons are released from the donor (e). This induces the PL suppression effect for the PL_2 and the PL enhancement effect for PL_1. At the same time, the BGE excites electrons up to the donor (f), which causes additional PL_2. What we can observe are the PL_1 and PL_2, which are collective of the above effects induced by the AGE and BGE. In order to examine the various effects separately, we chop the excitation source and detect the synchronized PL signal under the following five excitation modes; (i) $\overline{\text{AGE}}$, (ii) $\overline{\text{AGE}}+\overline{\text{BGE}}$, (iii) $\overline{\text{BGE}}$, (iv) $\overline{\text{AGE}}+\text{BGE}$, and (v) $\text{AGE}+\overline{\text{BGE}}$, where the sign of ‾‾ and —— denote the chopped and continuous excitation, respectively. In Fig. 15, the excitation intensities of the AGE and BGE and the resulting PL signals are illustrated for the above five excitation modes. For example, the PL_2 intensity under the $\text{AGE}+\overline{\text{BGE}}$ mode is t as shown in Fig. 15 (v), which is the superimposition of the AGE-PL with the suppression effect caused by the BGE (q) and the BGE-PL (s). The intensities p, q, s, $r (=s-q)$, and t are directly obtained as a synchronized signal.

The detailed analysis for a simpler system was reported by Monemar and Samuelson[60] and Grimmeiss and Monemar[61], although they neither counted the BGE-PL nor adopted chopped excitation technique. They investigated the PL suppression effect as a function of photon energy of the BGE, and thus obtained spectra of optical cross sections. In actual semiconductors, the

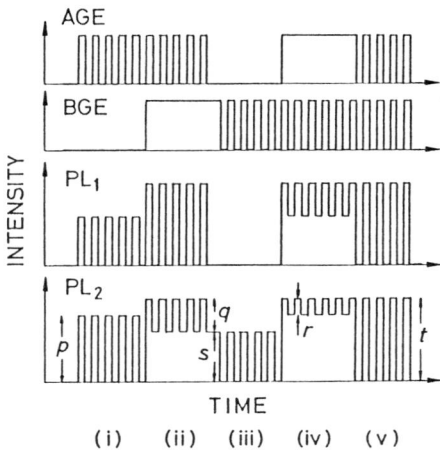

Fig. 15. Schematic illustration of AGE and BGE intensities and PL responses for band-to-band emission (PL_1) and donor-to-band emission (PL_2) under five excitation modes as a function of time. (Ref. 26)

existence of multiple levels, both radiative and nonradiative centers, makes it difficult to perform the quantitative analysis. However, the present PL analysis under the five excitation modes gives us a useful clue to investigate various recombination processes which could not be clarified otherwise. The present method is applicable to the assessment of nonradiative deep centers, since the PL enhancement effect for PL_1 takes place whether or not the recombination process via the donor level is radiative.

Samples used for the present study were undoped GaAs crystals grown by LEC and Horizontal Bridgeman (HB) methods. They were obtained from four different sources. We used the 514.5-nm line of an Ar laser as the AGE source and the 1.06-μm line of a YAG laser as the BGE source. The two laser beams hit the same spot on the sample at 4.2 K. The beam size was about 3 mm in diam. and the incident power was about 3 W·cm^{-2} for both excitations. The chopping frequency of the excitation beams was primarily 90 Hz.

PL spectra of the present samples consist of the 1.49-eV band and 0.6-eV band. The peak position of the 0.6-eV band varies from 0.62 to 0.68 eV depending on the samples and on the excitation conditions. In the following part we show the results on the sample #07 as a typical example; this sample is the same one as used in Sec. 6. As for the other samples, the data are summarized in Table 3.

The PL spectra of the 1.49-eV band taken under the $\overline{\text{Ar}}$, $\overline{\text{Ar}}+\overline{\text{YAG}}$, and $\overline{\text{Ar}+\text{YAG}}$ modes are shown in Fig. 16. A remarkable increase in the PL intensity is observed under twofold excitation. It should be noted that the PL intensity is higher under the $\overline{\text{Ar}+\text{YAG}}$ mode than under the $\overline{\text{Ar}}+\overline{\text{YAG}}$ mode. The simple model in Figs. 14 and 15 cannot account for this result. As we will discuss later the sample probably contains a high concentration of nonradiative deep centers. We consider it reasonable to assume that the continuous Ar laser excitation supplies more electrons (or holes) to the deep centers than the chopped Ar laser excitation. Then, the concentration of excess carriers excited by the BGE is higher in the $\overline{\text{Ar}}+\overline{\text{YAG}}$ mode than in the $\overline{\text{Ar}+\text{YAG}}$ mode. This explains the above PL enhancement effect.

The intensity maximum of the 1.49-eV band taken under twofold excitation occurs at a higher energy side than that under the single AGE. The 1.49-eV band has been reported to the due to band-to-acceptor transition (eA°) and donor-to-acceptor transition (D°A°).[33] The YAG laser light ionizes the neutral donors and acceptors responsible for the 1.49-eV band. Therefore, the intensity ratio of the D°A° emission to the eA° emission is reduced by the YAG laser irradiation. This causes the above spectral shift.

The 0.6-eV band can be excited either by the Ar laser irradiation or by the YAG laser irradiation, although the origins of the two kinds of PL are not necessarily the same (See Sec. 5 and 6). The 0.66-eV band excited by the Ar laser light is reduced by the YAG laser illumination as shown in Fig. 17(a). This is the suppression effect. On the other hand, the 0.68-eV band excited by

Table 3. PL enhancement and suppression effects under twofold excitation.

Sample	1.49-eV band					0.6-eV band				
	$\widetilde{Ar}$	$\overline{Ar+YAG}$	$\widetilde{YAG}$	$\overline{Ar+YAG}$	$\widetilde{Ar}+\widetilde{YAG}$	$\widetilde{Ar}$	$\overline{Ar+YAG}$	$\widetilde{YAG}$	$\widetilde{Ar}-\widetilde{YAG}$	$\widetilde{Ar}+\widetilde{YAG}$
LEC#65	1.0	3.1	0.0	4.3	3.6	1.0(0.65)	0.90(0.65)	0.20(0.65)	0.20(*)	1.1(0.65)
LEC#15	1.0	2.6	0.0	3.0	3.0	1.0(0.63)	1.1(0.63)	0.01(0.62)	0.12(0.63)	1.1(0.63)
LEC#07	1.0	2.1	0.0	2.8	2.6	1.0(0.66)	0.71(0.68)	1.0(0.68)	2.0(0.68)	2.5(0.68)
LEC#05	1.0	1.9	0.0	1.5	2.0	1.0(0.68)	1.0(0.68)	0.50(0.68)	0.50(0.68)	1.5(0.68)
LEC#66	1.0	1.7	0.0	1.7	2.0	1.0(0.64)	0.85(0.64)	0.03(0.65)	−0.08(*)	0.87(0.64)
LEC#60	1.0	1.4	0.0	0.68	1.5	1.0(0.68)	0.80(0.68)	1.6(0.68)	1.6(0.68)	2.8(0.68)
HB #27	1.0	1.0	0.0	0.0	1.0	1.0(0.63)	0.89(0.63)	0.29(0.65)	0.25(0.65)	1.1(0.63)
HB #20	1.0	0.87	0.0	−0.20	0.85	1.0(0.63)	1.0 (0.63)	1.1(0.63)	1.1(0.63)	2.1(0.63)

The values in the parentheses denote the peak position of the 0.6-eV band in eV (*; peak position cannot be defined because of the big spectral distortion in the 0.6-eV band).

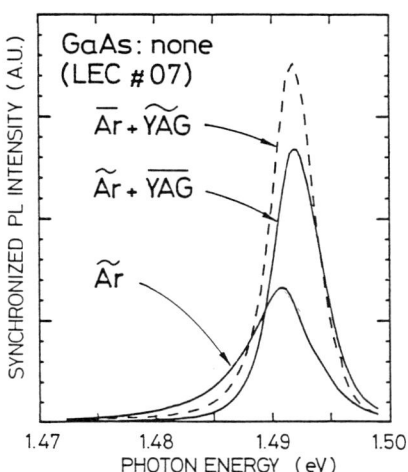

Fig. 16. PL spectra of 1.49-eV band in undoped LEC GaAs at 4.2 K under $\overline{Ar}$, $\widetilde{Ar}+\overline{YAG}$, and $\overline{Ar}+\widetilde{YAG}$ modes. (Ref. 26)

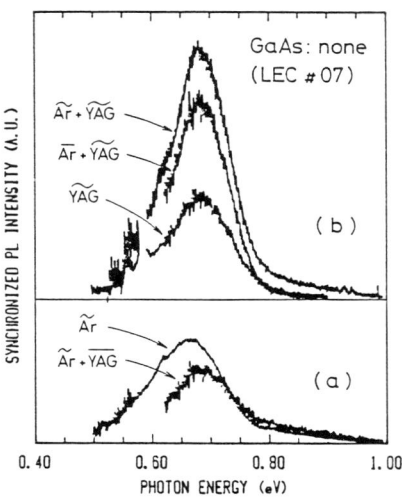

Fig. 17. PL spectra of 0.6-eV band in undoped LEC GaAs at 4.2 K under $\widetilde{Ar}$, $\widetilde{Ar}+\overline{YAG}$, $\widetilde{YAG}$, $\overline{Ar}+\widetilde{YAG}$, and $\widetilde{Ar}+\widetilde{YAG}$ modes.

the YAG laser light is enhanced by the Ar laser irradiation (Fig. 17(b)). The occurrence of the enhancement effect leads us to suggest that a high concentration of nonradiative deep centers should be present in addition to the deep centers responsible for the 0.6-eV band. This explains a drastic PL enhancement effect in the 1.49-eV band.

The PL enhancement and suppression effects under the five excitation modes for various samples are summarized in Table 3, where the relative PL intensities under various excitation modes with respect to the intensity under the $\widetilde{\text{Ar}}$ mode are listed for the 1.49- and 0.6-eV bands. The values in the parentheses denote the peak position of the 0.6-eV band in eV. As to the 1.49-eV band, a drastic PL enhancement effect is always observed for LEC samples but never for HB samples. At the same time most LEC samples show the suppression effect in the 0.6-eV band. However, the intensity reduction due to the suppression effect is 30% at most, which is too small to explain the drastic PL enhancement effect. This implies that a high concentration of nonradiative deep centers are present in addition to the deep levels responsible for the 0.6-eV band in LEC crystals but not in HB crystals. The factor of the PL enhancement is about 1.5 to 4 in the LEC samples. The concentration of carbon acceptors responsible for the 1.49-eV band is in the range of 5–10×10^{15} cm^{-3} obtained by infrared absorption. Therefore, the concentration of the nonradiative deep centers is roughly estimated at about 1×10^{16} cm^{-3}.

The LEC samples are classified according to the magnitude of the enhancement effect in the 1.49-eV band. The samples #65, #15 and #07 exhibit a very big enhancement effect. This suggests the existence of a considerable amount of nonradiative deep centers. The enhancement effect occurs also in the 0.6-eV band under the $\widetilde{\text{Ar}} + \widetilde{\text{YAG}}$ mode for the sample #15 and under the $\text{Ar} + \widetilde{\text{YAG}}$ mode for the sample #07, which favors the above idea. In the samples #05, #66 and #60, the enhancement effect in the 1.49-eV band is not so big. Correspondingly the enhancement effect does not take place in the 0.6-eV band.

Experiments on excitation intensity dependence of the PL under twofold excitation and on a variation of PL as a function of photon energy of the BGE are now undertaken.

In conclusion, the twofold excitation modulated PL technique has been proved to be very useful to assess the deep levels in undoped GaAs bulk crystals. The PL enhancement and suppression effects are induced by the additional BGE as a result of optical transitions of carriers in deep levels. As to the near band-edge emission, a drastic enhancement effect is always observed for LEC samples but never for HB samples. This effect is so big that it cannot be counterbalanced with a slight suppression in the deep level PL. This fact allows us to conclude that a high concentration of nonradiative deep centers are present in LEC crystals but not in HB crystals.

8. Summary

Deep level PL of undoped SI LEC GaAs crystals has been investigated in detail under various excitation conditions. The results are summarized as follows:

(1) The two typical bands, the 0.6-eV band and the 0.8-eV band, appear in the PL under the AGE condition. The appearance of the two bands depends strongly on the dislocation and the melt composition.

(2) The 0.6-eV band shows the PL fatigue and recovery effects, which are explained by the optical transitions between the normal state and the metastable state of the EL2 level.

(3) Wide variations in the peak position and the half-width and in the fatigue and recovery rates for the 0.6-eV band indicate that the origin of the EL2 level is not unique (multiple EL2 levels).

(4) The 0.67-eV band observed under the 1.32-μm light excitation is associated with one of the multiple EL2 levels. This level is commonly present in undoped SI LEC crystals.

(5) The 0.8-eV band is not associated with the EL2 levels but with the microdefects produced under the As-rich melt composition during the crystal growth.

(6) The twofold excitation modulated PL spectroscopy reveals nonradiative deep centers in LEC crystal. These centers are not present in boat-grown crystals.

We believe the present results will contribute greatly to a fundamental understanding of the complicated properties of the deep levels.

Acknowledgements

The author would like to thank K. Hoh and Y. Tokumaru for useful discussions and comments, Y. Okada for the data on etch pit/optical microscopy, and H. Tanino for helpful discussions on the configuration coordinate model. This work was partly supported by Special Coordination Funds for Promoting Science and Technology.

REFERENCES

1) G. M. Martin, J. P. Farges, G. Jacob, and J. P. Hallais: J. Appl. Phys. **51** (1980) 2840.
2) E. J. Johnson, J. A. Kafalas, and R. W. Davies: J. Appl. Phys. **54** (1983) 204.
3) For example, R. D. Fairman, R. T. Chen, J. R. Oliver, and D. R. Ch'en: IEEE Trans. Electron Devices **ED-28** (1981) 135.
4) For review articles, S. Makram-Ebeid, P. Langlade, and G. M. Martin: *Semi-Insulating III-V Materials, Kah-nee-ta, 1984*, ed. D. C. Look and J. S. Blakemore (Shiva, Nantwich, 1984) p. 184; J. Lagowski and H. C. Gatos: *Proc. 13th Int. Conf. Defects in Semiconductors*, ed. L. C. Kimerling and J. M. Parsey, Jr. (AIME, Warrendale, 1985) p. 73.
5) G. M. Martin: *Semi-Insulating III-V Materials, Nottingham 1980*, ed. G. J. Rees (Shiva, Orpington, 1980) p. 13.

6) G. M. Martin: Appl. Phys. Lett. **39** (1981) 747.
7) D. E. Holmes, R. T. Chen, and J. Yang: Appl. Phys. Lett. **42** (1983) 419; D. E. Holmes and R. T. Chen: J. Appl. Phys. **55** (1984) 3588.
8) M. R. Brozel, I. Grant, R. M. Ware, D. J. Stirland, and M. S. Skolnick: J. Appl. Phys. **56** (1984) 1109.
9) W. J. Turner, G. D. Pettit, and N. G. Ainslie: J. Appl. Phys. **34** (1963) 3274.
10) B. Deveaud and P. N. Favennec: *Gallium Arsenide and Related Compounds 1978*, ed. C. M. Wolfe (Inst. Phys., Bristol, 1979) p. 492.
11) P. W. Yu: Solid State Commun. **32** (1979) 1111.
12) A. Mircea-Roussel and S. Makram-Ebeid: Appl. Phys. Lett. **38** (1981) 1007.
13) P. W. Yu, D. E. Holmes, and R. T. Chen: *Gallium Arsenide and Related Compounds 1981*, ed. T. Sugano (Institute of Physics, Bristol, 1982) Conf. Ser. 63, p. 209.
14) M. Tajima: Jpn. J. Appl. Phys. **21** (1982) L227.
15) P. Leyral, G. Vincent, A. Nouailhat, and G. Guillot: Solid St. Commun. **42** (1982) 67.
16) J. Windscheif, H. Ennen, U. Kaufmann, J. Schneider, and T. Kimura: Appl. Phys. **A30** (1983) 47.
17) M. Tajima and Y. Okada: *Defects in Semiconductors*, ed. C. A. J. Ammerlaan (North-Holland, Amsterdam, 1983) Physica **B 116**, 404.
18) B. V. Shanabrook, P. B. Klein, E. M. Swiggard, and S. G. Bishop: J. Appl. Phys. **54** (1983) 336.
19) B. V. Shanabrook, P. B. Klein, and S. G. Bishop: Physica **117B & 118B** (1983) 173.
20) P. W. Yu and D. C. Walters: Appl. Phys. Lett. **41** (1982) 863.
21) D. J. Wolford, S. Modesti, and B. G. Streetman: *Gallium Arsenide and Related Compounds 1982*, ed. G. Stillman (Institute of Physics, Bristol, 1983) Conf. Ser. 65, p. 501.
22) T. Kikuta, K. Terashima, and K. Ishida: Jpn. J. Appl. Phys. **22** (1983) L409.
23) P. W. Yu: Appl. Phys. Lett. **44** (1984) 330.
24) M. Tajima: Jpn. J. Appl. Phys. **23** (1984) L691.
25) D. Pajet and P. B. Klein: *Proc. 13th Int. Conf. Defects in Semiconductors*, ed. L. C. Kimerling and J. M. Parsey, Jr. (AIME, Warrendale, 1985) p. 959.
26) M. Tajima: *Proc. 13th Int. Conf. Defects in Semiconductors*, ed. L. C. Kimerling and J. M. Parsey, Jr. (AIME, Warrendale, 1985) p. 997.
27) T. Kikuta, H. Emori, T. Fukuda, and K. Ishida: *Extended Abstracts of 16th Conf. Solid State Devices and Materials, Kobe, 1984*, p. 173.
28) L. Samuelson, P. Omling, and H. G. Grimmeis: Appl. Phys. Lett. **45** (1984) 521.
29) M. Tajima: Jpn. J. Appl. Phys. **24** (1985) L47.
30) M. Tajima: Appl. Phys. Lett. **46** (1985) 484.
31) M. Tajima: *Semiconductor Technologies 1982*, ed. J. Nishizawa (North-Holland, Amsterdam, 1982) p. 1.
32) P. J. Dean: *Progress in Crystal Growth Characterization 1982* (Pergamon Press, Oxford, 1982) p. 89.
33) D. J. Ashen, P. J. Dean, D. T. J. Hurle, J. B. Mullin, A. M. White, and P. D. Greene: J. Phys. Chem. Solids **36** (1975) 1041.
34) G. M. Martin, A. Mitonneau, and A. Mircea: Electron. Lett. **13** (1977) 191.
35) A. M. Huber, N. T. Linh, M. Valladon, J. L. Debrun, G. M. Martin, A. Mitonneau, and A. Mircea: J. Appl. Phys. **50** (1979) 4022.
36) D. E. Holmes, K. R. Elliott, R. T. Chen, and C. G. Kirkpatrick: *Semi-Insulating III-V Materials, Evian, 1982*, ed. S. Makram-Ebeid and B. Tuck (Shiva, Nantwich, 1982) p. 19.
37) P. W. Yu: Solid State Commun. **43** (1982) 953.
38) G. Vincent and D. Bois: Solid St. Commun. **27** (1978) 431.
39) G. Vincent, D. Bois, and A. Chantre: J. Appl. Phys. **53** (1982) 3643.
40) M. S. Skolnick, L. J. Reed, and A. D. Pitt: Appl. Phys. Lett. **44** (1984) 447.

41) L. Samuelson, P. Omling, E. R. Weber, and H. G. Grimmeiss: *Semi-Insulating III-V Materials, Kah-nee-ta, 1984*, ed. D. C. Look and J. S. Blakemore (Shiva, Nantwich, 1984) p. 268.
42) E. W. Williams and H. B. Bebb: *Semiconductors and Semimetals*, ed. R. K. Willardson and A. C. Beer (Academic, NY, 1972) Vol. 8, p. 370.
43) P. Leyral and G. Guillot: *Semi-Insulating III-V Materials, Evian, 1982*, ed. S. Makram-Ebeid and B. Tuck (Shiva, Cheshire, 1982) p. 166.
44) Y. Okada: Jpn. J. Appl. Phys. **22** (1983) 413.
45) A. G. Cullis, P. D. Augustus, and D. J. Stirland: J. Appl. Phys. **51** (1980) 2556.
46) A. G. Tweet: J. Appl. Phys. **30** (1959) 2002.
47) T. Iizuka: J. Electrochem. Soc. **118** (1971) 1190.
48) A. J. R. de Kock: *Physics of Semiconductors 1978*, ed. B. L. H. Wilson (Inst. Phys., Bristol, 1979) p. 103.
49) M. Tajima, S. Kishino, M. Kanamori, and T. Iizuka: J. Appl. Phys. **51** (1980) 2247.
50) H. C. Casey, Jr.: J. Electrochem. Soc. **114** (1967) 153.
51) G. A. Rozgonyi and M. A. Afromowitz: Appl. Phys. Lett. **19** (1971) 153.
52) T. Kikuta, T. Terashima, and K. Ishida: Jpn. J. Appl. Phys. **22** (1983) L541.
53) K. Kuramoto (private communication): orally presented at the 44th Autum Meeting of the Japan Society of Applied Physics, Sendai, September, 1983.
54) S. Makram-Ebeid, P. Langlade, and G. M. Martin: *Semi-Insulating III-V Materials, Kah-nee-ta, 1984*, ed. D. C. Look and J. S. Blakemore (Shiva, Nantwich, 1984) p. 184.
55) A. Mitonneau and A. Mircea: Solid St. Commun. **30** (1979) 157.
56) M. Taniguchi and T. Ikoma: J. Appl. Phys. **54** (1983) 6448.
57) M. Taniguchi and T. Ikoma: Appl. Phys. Lett. **45** (1984) 69.
58) I. Fujimoto: Jpn. J. Appl. Phys. **23** (1984) L287.
59) F. A. Ponce, F-C. Wang, and R. Hiskes: *Semi-Insulating III-V Materials, Kah-nee-ta, 1984*, ed. D. C. Look and J. S. Blakemore (Shiva, Nantwich, 1984) p. 68.
60) B. Monemar and L. Samuelson: Phys. Rev. **18** (1978) 809.
61) H. G. Grimmeiss and B. Monemar: Phys. Stat. Sol. (a) **19** (1973) 505.
62) A. Steckenborn, H. Münzel, and D. Bimberg: J. Lumin. **24/25** (1981) 351.
63) T. Kamejima, F. Shimura, Y. Matsumoto, H. Watanabe, and J. Matsui: Jpn. J. Appl. Phys. **21** (1982) L721.
64) A. K. Chin, A. R. Von Neida, and R. Caruso: J. Electrochem. Soc. **129** (1982) 2386.
65) B. Wakefield, P. A. Leigh, M. H. Lyons, and C. R. Elliott: Appl. Phys. Lett. **45** (1984) 66.
66) A. K. Chin, R. Caruso, M. S. S. Young, and A. R. Von Neida: Appl. Phys. Lett. **45** (1984) 552.
67) S. Miyazawa and Y. Nanishi: *Proc. 14th Conf. (1982 Int.) Solid State Devices, Tokyo, 1982*, Jpn. J. Appl. Phys. Suppl. **22-1** (1983) 419.
68) A. S. Jordan, R. Caruso, and A. R. Von Neida: Bell Syst. Tech. J. **59** (1980) 593.

Defects and Properties of Semiconductors: Defect Engineering, edited by J. Chikawa, K. Sumino, and K. Wada, pp. 71–86.
© KTK Scientific Publishers, Tokyo, 1987.

ANALYSIS OF NONSTOICHIOMETRY AND DOPED IMPURITIES IN GaAs BY X-RAY QUASI-FORBIDDEN REFLECTION (XFR) METHOD

Isao FUJIMOTO

Science and Technical Research Laboratories of NHK, 1-10-11 Kinuta, Setagaya-ku, Tokyo 157, Japan

Abstract. Nonstoichiometry and doped impurities in GaAs were investigated by measuring the intensities of quasi-forbidden reflections of X-rays. Most of Si atoms in MBE growth at high doping level were found to occupy Ga sites, contrary to the case of LPE growth where they occupy both Ga and As sites with almost equal probabilities. Other impurities like Al, In and Zn were also analyzed. Fairly large deviations ($\sim 10^{-4}$) with higher atomic concentrations in As lattice plane were found for liquid-encapsulated Czochralski (LEC) grown crystals compared with horizontal Bridgman (HB) grown ones, and their distribution were observed to have a close correlation with EPD. Nonstoichiometry of epitaxial layers was found to depend on the growth condition.

1. Introduction

It has recently become evident that nonstoichiometry has a large influence on the electrical, optical and mechanical properties of III-V compound semiconductors, *e.g.*, for GaAs, change of carrier concentration due to heat treatment under various As pressure,[1] effect of As pressure on the perfection of crystals grown by a horizontal Bridgman method,[2] a large dependence of the charactersitics of LEC grown crystals on the composition of the melt,[3,4] thermal conversion during heat treatment and insulator-semiconductor interface problems. The low reliability and yield of GaAs devices compared with Si are due mainly to these unstable composition of compound semiconductors. Therefore, the characterisation of nonstoichiometry has received an increasing attention and various approaches, both theoretical and experimental, have been made; calculation of the existence region of solid GaAs in phase diagram by thermodynamical analysis,[5] lattice parameter measurement,[6,7] coulometric titration, ion beam scattering,[8] photoluminescence measurement and other approaches. However, there has been no straightforward method of its characterisation with enough accuracy and the detail of nonstoichiometry is not fully clarified yet. In the present work, a new method is presented by which a deviation from stoichiometry can be detected with an accuracy as high as $c_{As}-c_{Ga}\sim 3\times 10^{-5}$.[9,10] This method can also be applied to the determination of the lattice location of doped impurities as well as their concentration at the doping level over 10^{18} cm^{-3}. Especially, the XFR method is very useful for the investigation of the behaviour of amphoteric impurities such as Si in GaAs, whose definite picture has not been obtained yet.

First, the principle of the XFR method is shown with brief description of the experimental procedures. Next, the capabilities of the method is demonstrated by the study of doped impurities. Then the analysis of nonstoichiometry of crystals grown by various methods and the effect of heat treatment on it are presented. Finally, some problems to be solved and the scope of future applications are briefly discussed.

2. Principle of the Measurement

The intensities of X-rays diffracted by a crystal are specified by the structure factor F which reflects the configuration of the constituent atoms as follows:

$$F_h = \sum_j (f^0{}_j + f'_j + if''_j) \exp(-M_j) \exp(2\pi i h \cdot r_j) \qquad (1)$$

where $f^0{}_j$ is the atomic scattering factor of a j-th atom, f'_j and f''_j are the anomalous dispersion correction, $\exp(-M_j)$ is Debye-Waller factor and r_j is the atomic coordinate. There are some weak reflections in which X-rays scattered from Ga and As atoms are in opposite phases. For example, the hkl reflections with $h+k+l=4n+2(n=0, 1, 2, —)$ arise from the small difference in scattering factors between Ga and As:

$$F = 4(f_{Ga} - f_{As}) \qquad (2)$$

where $\qquad f_j = (f^0{}_j + f'_j + if''_j)\exp(-M_j).$

Equation (2) is the ideal formula assuming that Ga and As sites are fully occupied by the respective atom. In real crystals, nonstoichiometry is expected; there exist native defects such as vacancies, interstitials and their complexes which make the probability of finding Ga and As atom at the respective site differ from unity. Since atomic scattering power increase with the number of electrons of the atom, vacancies and/or substitutional impurities with the atomic number much smaller than the host atoms diminish

Analysis of Nonstoichiometry and Doped Impurities in GaAs

the scattering power of the net plane, whereas interstitials and/or impurities with larger atomic numbers increase it. It should be noted here that the effect of antisite is small because the atomic numbers of Ga and As are very close to each other. Then, eq. (2) should be modified as

$$F = 4(c_{Ga}f_{Ga} - c_{As}f_{As}). \qquad (3)$$

Here, c_j is an atomic concentration ratio in a respective net plane relative to the ideal crystal ($c_{Ga} = c_{As} = 1$ for ideal crystal), and given as follows;

$$c_A = \{[A_A] + [A_I] + (f_x/f_A)([X_A] + [X_I])\}/N_A$$

$$= 1 + \{[A_i] - [V_A] + (f_x - f_A)(1/f_A)[X_A] + (f_x/f_A)[X_I]\}/N_A. \qquad (4)$$

where
- A_A: A atoms at regular sites of A.
- V_A: A vacanicies.
- A_I: self-interstitials in A atomic plane (tetrahedral sites).
- X_A: substitutional impurities in A atomic plane.
- X_I: interstitial impurities in A atomic plane.
- N_A: number of A sites per unit volume.
- f_A, f_x: scattering factors of respective atoms, A and X.

A schematic presentation of eq. (3) in a complex structure factor diagram is given in Fig. 1. It can readily be seen that a small deviation from unity in $c_{A,B}$

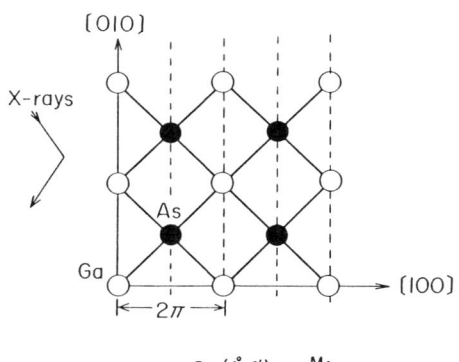
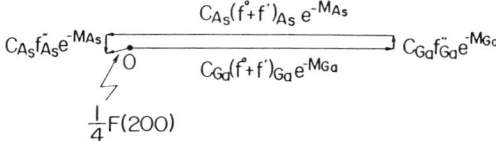

Fig. 1. Schematic presentation of the structure factor of 200 reflection.

results in a large variation of diffracted intensities of X-rays; $I(200)$ decreases for $c_{Ga} > c_{As}$ and increases for $c_{Ga} < c_{As}$. Therefore, by measuring the integrated intensities of the forbidden reflections, we can determine the deviation from stoichiometry and lattice locations of impurities. For the absolute determination of nonstoichiometry, however, accurate values of structure factors are required in addition to the absolute measurements of intensities. At present, we have not the definite values of $F(200)$ with enough accuracy. Therefore, in the present study, nonstoichiometry was measured relatively to a horizontal Bridgman (HB) grown crystal which is considered to have stable quality because of growth under equilibrium conditions. Thus we have

$$\Delta I / I = K(c_{As} - c_{Ga}) \tag{5}$$

where ΔI and $(c_{As} - c_{Ga})$ means those values relative to the HB grown reference crystals. The conversion factor, K, depends on the reflection indices and the radiations used. For 200 reflection, for example, an intensity variation of 0.1% corresponds to a small deviation of $c_{As} - c_{Ga} = 3 \times 10^{-5}$ and 3.7×10^{-5} for MoKα and CuKα radiation, respectively. For 600 reflection, the intensity variations are smaller than those of 200 reflection by a factor of 0.4 for the same amount of deviation.

It should be noted here that in the above formulation for $h00$ ($h = 2, 6, —$) reflection all atoms are assumed to be in the regular lattice plane of Ga and As. Those atoms at lattice sites other than the regular sites, such as interstitials at bond center or at hexagonal symmetry sites, interstitials of dumb-bell type and those displaced atoms around interstitials and impurities, should be taken into account with the appropriate phase factor of $\exp(2\pi i h \cdot r_j)$. Especially, it is worth noting that the effect of atomic displacement is much higher for higher indexed reflections than for lower indexed one; the effect for 600 reflection is about 10 times that for 200 reflection.

Thus we can determine the concentrations and microscopic atomic structure of nonstoichiometric defects and impurities by measuring the relative intensities of various quasiforbidden reflections of X-rays.

On the basis of diffraction theory, the diffracted intensities of X-rays depend on the perfection of crystals; generally, lattice distortion increases the diffracted intensities of X-rays, and those defects like dislocations can be observed in X-ray topographs. However, for such a weak reflections as the quasi-forbidden reflections the effect is substantially small, e.g., the ratios of I(mosaic)/I(perfect) for 200 reflection were calculated to be 1.10, 1.01 and 1.03 for CuKα, MoKα and AgKα radiations, respectively, whereas the ratios for 400 reflection were 7.0, 3.9 and 5.5, respectively. This was confirmed experimentally by measuring the change in intensities due to surface damage caused by lapping; $I(200)$ was found to increase only by less than 1.8%, whereas $I(400)$ to increase by more than 25% for CuKα. Actually, the integrated intensities of 400 reflection from a highly dislocated region (EPD: $\sim 1\times 10^5$ cm^{-2}) was observed to be about 4% higher than that from a perfect region. Therefore, the dislocation effect on the intensities of a quasi-forbidden reflection is considered to be negligibly small. The strain effect is still smaller with use of MoKα radiation.

3. Experimental Procedure

MoKα and CuKα radiations from a high power X-ray generator (60 kV 500 mA max.) were monochromatised by Ge 111 reflection. The integrated intensities were measured by rotating the specimen through Bragg position by an angle of 4′–20′ of arc. Since for such a weak reflection, the effect of Umweganregung is quite large, the specimen was rotated around the axis vertical to the reflecting plane and set to an Umweganregung-free position. Measurements were made in pairs by rotating the specimen 180° around this axis to make a small correction for a misorientation between the crystal surface and the diffracting plane. The intensities of X-rays incident on specimen were monitored by measuring the scatterd X-rays from a thin mylar film.

4. Results

4.1 Impurity effect

The effect of doped impurities on the integrated intensities of quasiforbidden reflections were investigated. Specimens doped with Si and Al were prepared by molecular beam epitaxy (MBE) at substrate temperature $T_s=600°$ C, $J_{As_4}/J_{Ga}=1.4\sim 5.0$ and growth rate of $1.3\sim 2.6\,\mu$m/h. In addition, two highly Si doped LPE wafers, one was n-type grown at $930\sim 920°$C and the other was p-type grown at $700\sim 650°$C, were prepared for comparison with the MBE grown one. Those doped with In were grown by LEC technique. The Zn doped one was prepared by diffusion of Zn using ZnAs$_2$ at 700°C for 4 hours. Concentration of doped impurities was determined by SIMS analysis. The results are shown in Figs. 2–5. It should be noticed that the ratio of the intensity changes using CuKα to those by MoKα (~ 0.8) agreed with the calculated value so that the observed values by MoKα are plotted in values reduced to CuKα one.

4.1.1 Al and In

As can be expected, the intensity of 200 reflection increases when Al, which has smaller atomic number than host atoms, occupy Ga site, whereas it decreases when heavier atoms like In occupy Ga site. The agreement between the measured values and the calculated one is satisfactory (Fig. 2). Moreover, good agreement between theory and experiment obtained for 600 reflection, for which $\Delta I/I$ is much sensitive to atomic displacement, implies that Al

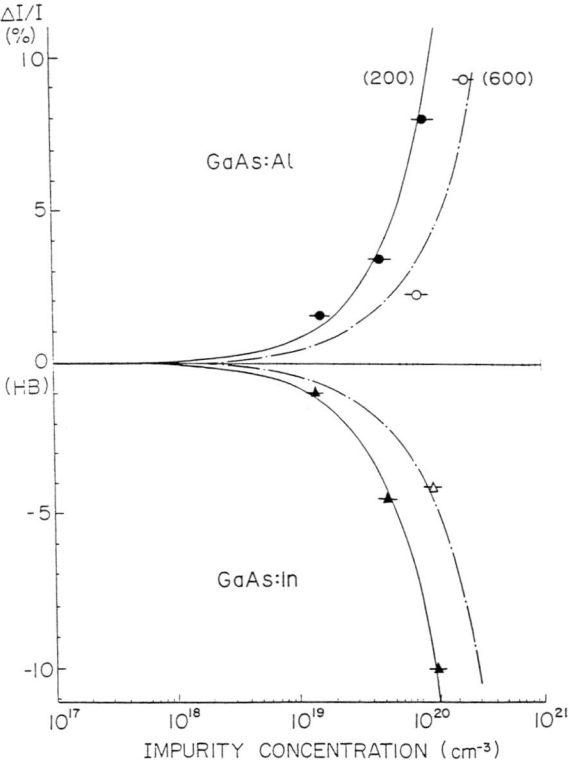

Fig. 2. Intensity variations of XFR vs concentration of doped impurities of Al and In. Solid and dotted lines are the calculated curves for 200 and 600 reflections, respectively, assuming Al_{Ga} and In_{Ga}.

occupy substitutional sites with small amount of lattice distortion. It should be noted that distribution of impurity concentration can be visualized by taking a topography of XFR. One example is shown in Fig. 3 for In doped LEC wafers. Striations due to fluctuation of In concentration can be clearly seen. Remarkable thing is that those images of black spots in the topograph by 400 reflection, where intensities are high because of lattice distortion, reverse contrast in 200 reflection topograph because In segregation has reduced the intensities.

4.1.2 Zn

For a highly Zn doped specimen with densities of the order of 5×10^{20} cm^{-3} at surface, and 1×10^{20} cm^{-3} at the diffusion front of about 10 μm, a fairly large decrease in the intensities was observed for 200 reflections as shown in

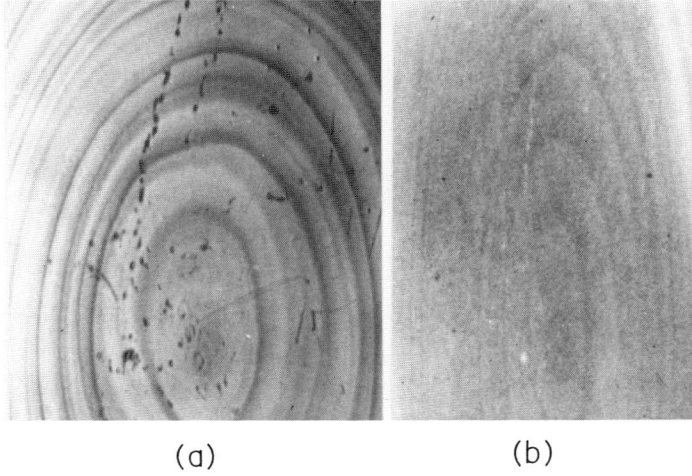

Fig. 3. Reflection X-ray topographs of In doped LEC wafer. (a) 400 reflection, (b) 200 reflection, CuKα. Note the reverse contrast of spotty images in (a) and (b).

Fig. 4, while almost no decrease was observed for 600 reflection. The difference in atomic number between Zn and Ga is so small that substitutional Zn atoms contribute only a small fraction to the change in XFR intensities. Therefore, the decrease in the intensities of 200 reflection can be explained by assuming a high concentration of Zn ($\sim 5\times 10^{19}$ cm^{-3}) at tetrahedral interstitial sites surrounded by As atoms. The deviation of the observed intensity of 600 reflection from the calculated one on the basis of 200 reflection implies the effect of lattice displacement as described above, possibly slight displacement (~ 0.1 Å) of Zn from the regular interstitial sites. The observation of a high concentration of Zn_I agree with the results reported by other authors.[12] These Zn_I may be responsible for the disordering of superlattices induced by Zn diffusion.

4.1.3 Si

In the analysis of doped impurities in GaAs using XFR method, most remarkable result has been obtained for the case of Si doping. Si has been widely used as an excellent n-dopant in GaAs. However, because of its amphoteric nature, the properties of the material depend very much on growth conditions. Therefore, doping of Si into GaAs has been the subject of a number of investigations:[13,14] In MBE growth, for example, the carrier concentration saturates or even decreases at a high doping level over 5×10^{18} cm^{-3}. This has been attributed to amphoteric nature of Si; at low doping level, Si atoms occupy Ga sites and act as donors, while at high doping level they

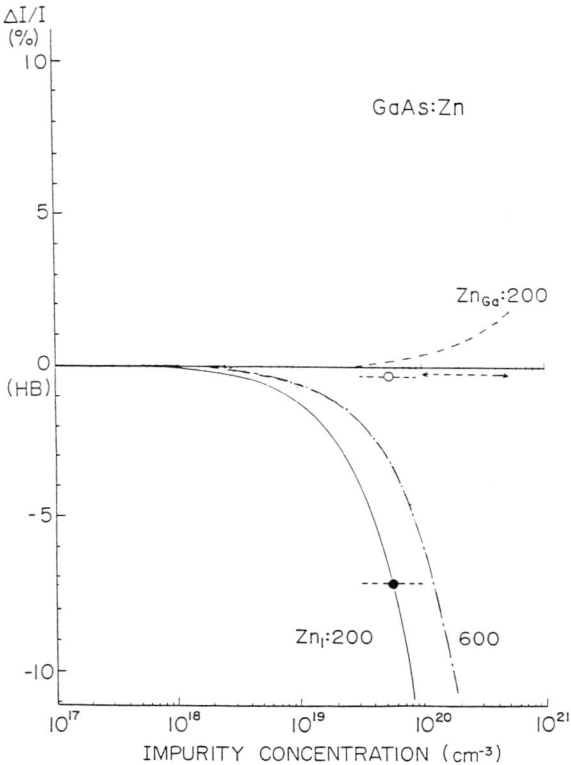

Fig. 4. Intensity variations of XFR vs Zn concentration. Calculated curves are for Zn_{Ga} (200 reflection) and for Zn_I surrounded by As atoms (200 and 600 reflection). Net concentration of Zn is $1\sim5\times 10^{20}$ cm^{-3}.

tend to occupy As sites and act as acceptors compensating Si_{Ga} donors. In LPE growth at high doping of Si, Si_{Ga}-Si_{As} pairs have been observed by infrared absorption spectra[15] and these pairs have been considered to be responsible for various properties such as rather high diffusion coefficient of Si in GaAs at high doping level.[16]

If this autocompensation mechanism dominates, the observed intensity variations of the quasi-forbidden reflections should saturate in a highly doped region over 5×10^{18} cm^{-3} as shown in Fig. 5 with a broken line. Contrary to this expectation, the intensity variations of 200 reflection increase linearly with doping concentration of Si up to 2×10^{20} cm^{-3} and coincides with the expected values assuming most of Si atoms to occupy Ga sites. At the doping level higher than 2×10^{20} cm^{-3}, $\Delta I/I$ deviate from the caluculated curve and starts

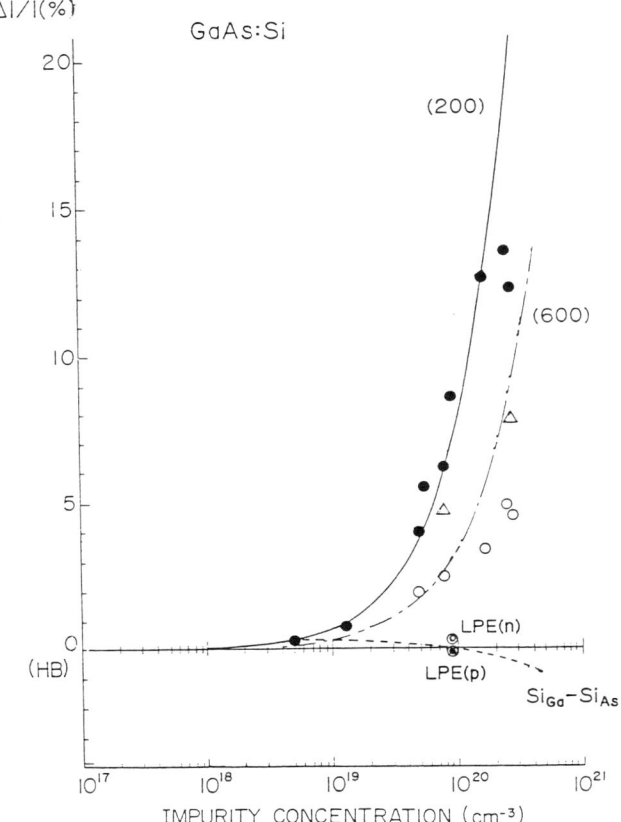

Fig. 5. Intensity variations of XFR vs Si concentration. The solid and the dotted lines show the calculated curves for 200 and 600 reflections, respectively, assuming Si_{Ga}. The broken line represents the calculated curve assuming autocompensation mechanism. Also shown are the change in $\Delta I/I$ after heat treatment at 850°C for $2h(\Delta)$.

leveling. The intensity changes of 600 reflection observed for these specimens were found to coincide with those values expected from the data for 200 reflection, which implies that atomic displacement is very small. On the other hand, two LPE grown crystals with total Si concentration of 9×10^{19} cm^{-3}, showed different behaviour; the observed intensities of 200 reflection did not show much difference from that of HB grown reference crystals. It was slightly higher for n-type specimen and slightly lower for p-type one and can well be explained by the autocompensation mechanism described above.

The remarkable behaviour of MBE grown crystals may be interpretted in a several ways:

(1) Most Si atoms occupy Ga sites.

(2) Autocompensation mechanism dominates as has been reported by many investigators. At the same time, high concentration of self-interstitials in As atomic plane are induced.

(3) Most Si atoms occupy tetrahedral interstitial sites in As atomic plane.

Among three possibilities, the third case may be excluded considering that PIXE (Particle Induced X-ray Emission) measurement did not detect such a high concentration of Si interstitials.[14] The second case may also be excluded because the idea that such a high concentration of native defects is induced is physically unreasonable. Moreover, it is difficult to explain the observed relation between ΔI and concentration of Si.

Thus the first case is considered to be most probable. The saturation or even a decrease of carrier density may be explained by assuming clusters of Si which do not act as donors. For example, four Si_{Ga} which are next nearest neighbour to each other and form a tetrahedron, or adding one more Si at the center of the tetrahedron may be a candidate. Clusters of larger size can, of course, explain the observed results. However, since these cluster are so small and coherent with lattice that they could not be detected by Transmission Electron Microscope, as many authors have reported, though the rocking curve of X-ray diffraction is broad and perfection is rather poor. These clusters are considered to scatter electrons and to act as nonradiative centers and may be the origin of the substantial reduction in mobilities and photoluminescence intensities at high doping level.

It is worth noting here that the observed behaviour of the lattice location of Si was not found to depend on the flux ratio γ in the range $1.4<\gamma<5$, though incorporation ratio of Si was found to increases with γ. On the other hand, when they are annealed (face to face in N_2 ambient gas at $700\sim850°C$ for 2 h), the intensities of XFR decreases substantially as shown in Fig. 5, which implies that at higher temperature Si tend to occupy As sites. With increasing the annealing time, however, the intensity does not change any more and the value reached is still much higher than that expected from autocompensation mechanism.

What makes the different behaviour of Si dopant between LPE and MBE growth? Since p-type LPE specimen was grown at temperatures around $650°C\sim700°C$, the difference may not be attributed to growth temperature alone. The main difference may arise from the growth condition; since LPE specimens are grown in Ga rich solution, doped Si atoms are more likely to occupy As sites than in MBE growth where As overpressure is normally used. Thermal equilibrium character of LPE growth and nonequilibrium character of MBE may be another explanation for the observed difference.

Thus the X-ray quasi-forbidden reflection method was proved to be a simple and easy ways of dopant analysis with the information of its occupant site.

4.2 Evaporation of As by Heat Treatment

(100) and (111) semi-insulating HB grown wafers were prepared and their surfaces half-coated by rf-sputtered SiO_2 thin film. Heat treatment was made in N_2 ambient gas at 850~890°C for 20~50 min. The results are shown in Table 1. The X-ray intensities of ordinary reflections increased for the uncoated parts because of the lattice distortion around the induced defects as can be seen in X-ray topograph (Fig. 6), whereas those for quasi-forbidden reflections decreased markedly. It can readily be seen on the basis of formula (5) that a fairly large amount of As is exhausted from the uncoated surfaces. For the (111) Ga surface, no difference was observed in the diffracted intensities of X-rays, indicating that the Ga surface is very stable for heat treatment. It should be noted that the observed deviations are the values averaged over the penetration depth of X-rays of a few μm from the surface.

Table 1. Nonstoichiometry of GaAs (HB) caused by heat treatment in N_2 gas (MoKα).

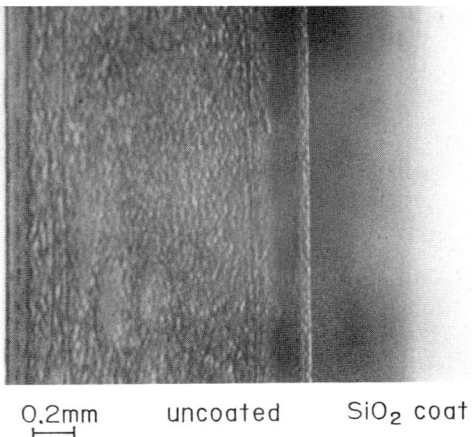

	(100) 850°C 50min.		(111) 890°C 20 min.			
hkl	200	400	222	333	$\bar{2}\bar{2}\bar{2}$	$\bar{3}\bar{3}\bar{3}$
I(A) / I(B)	0.78	1.28	1.00	1.00	0.86	1.16
$C_{Ga} - C_{As}$	5×10^{-3}		0		3.5×10^{-3}	

Fig. 6. Reflection X-ray topograph of the heat treated (100) wafer. (400) reflection.

4.3 LEC Grown Crystals

Extensive studies made on LEC grown crystals for the development of GaAs IC's have clarified many distinctive features compared with HB grown ones: Resistivities, mobilities, photoluminescence intensities and threshold voltages of FET was found to have W- or M-shaped profiles along a wafer diameter.[17,18] Since the growth condition such as the temperature gradient around the liquid to solid interface in LEC method is not so moderate as in HB method, a high concentration of native defects may still remain at room temperature. From this viewpoint, the undoped semi-insulating LEC grown crystals were investigated and has been reported previously; The observed 200 intensities showed W-shaped radial distribution and had a close correlation with EPD (etch-pit density) (Fig.7). The intensities higher than that of HB grown crystals indicated that atomic concentration of As atomic plane was

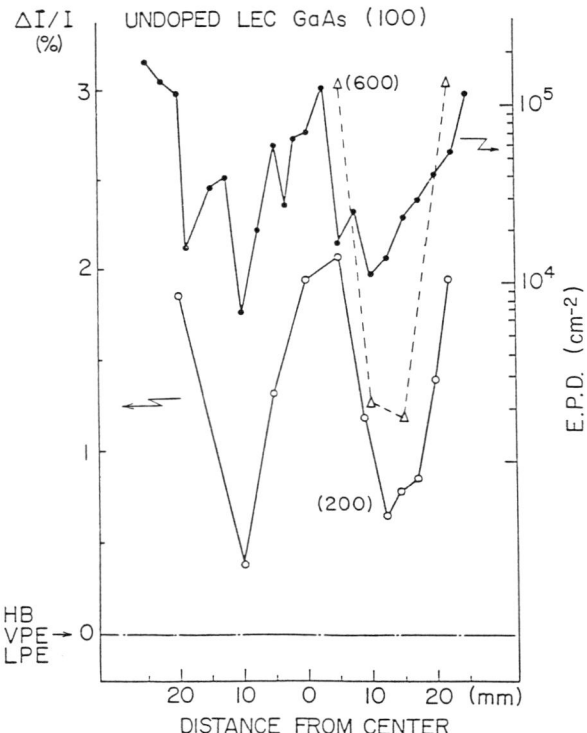

Fig. 7. Profiles of XFR intensities and EPD ditribution along wafer diameter of unoped LEC GaAs (100) wafer. Vertical axis shows the ratio of variations of XFR intensities relative to that of HB grown reference crystals (CuKα).

higher than that of Ga. It was deduced from this observation that interstitials at tetrahedral sites surrounded by Ga atoms might be the main defects related with nonstoichiometry. The observed high density of the order $10^{18} \sim 10^{19}$ cm^{-3} could be explained by assuming the clustere of such defects. The reported observations of As precipitates were considered to support this model.

In the present study detailed study was made on various LEC crystals using various radiations (CuKα and MoKα) and various reflections. The results vary from specimen to specimen depending on their growth conditions or heat treatment. Especially it is worth noting that the observed nonstoichiometry did not reflect the melt composition in a straightforward way; the XFR intensities of the specimen grown from Ga rich melt was found to be higher than that of HB grown crystals. Moreover, the theoretical relationships between $\Delta I/I$ (200) and $\Delta I/I$ (600) and between $\Delta I/I$ (CuKα) and $\Delta I/I$ (MoKα) does not hold anymore for LEC crystals, contrary to the case of impurity doping; $\Delta I/I$ (600)$>\Delta I/I$ (200) and $\Delta I/I$ (CuKα)$>\Delta I/I$ (MoKα). Therefore, simple model of As$_\mathrm{I}$ or their cluster alone proposed previously can not explain the results and real microscopic structure of the nonstoichiometric defects in LEC crystals is left unsolved at present, and more detailed study is necessary for its thorough understanding.

It should be noticed that face to face heat treatment at 850°C for 2 h in N$_2$ ambient gas reduces the deviation; intensities of XFR decrease and come close to those of HB reference crystals while HB crystals which was facing to the LEC specimen does not show any change. After etching the surface by 2~4 μm, the intensity goes back to the original value (Fig. 8). This observation may have a close relation with the reported outdiffusion of EL2 trap due to heat treatment.[19]

4.4 *MBE (molecular beam epitaxy) grown layers*

Extensive studies have shown that the characteristics of MBE grown GaAs layer varies with the flux ratio $\gamma = J_{As_4}/J_{Ga}$. Therefore, the present method was applied to the characterisation of nonstoichiometry of epitaxial layer (4 μm thick, Si doped: 1.5×10^{17}) grown under various γ (0.5<γ<5.6) by changing As$_4$ flux for the fixed Ga flux at the growth rate of 1.3 μm/h. Photoluminescence (PL) intensity at room temperature was also measured by using 514.5 nm Ar laser excitation. Since carrier concentration was found to be almost constant for γ>0.7 by the Hall measurements using Van der Pauw technique, PL intensity at room temperature is considered to depend on the concentration of nonradiative centers. The results are shown in Fig. 9a, b. For higher γ case (γ>2.8), the atomic concentration of As lattice plane is higher than that of Ga plane, and the PL intensity is weak. For 1<γ<2, the nonstoichiometry come close to that of HB crystals, and PL intensity was found to show maximum. For Ga rich condition (γ<0.7), however, the concentration of As lattice plane was also found to rise, and correspondingly

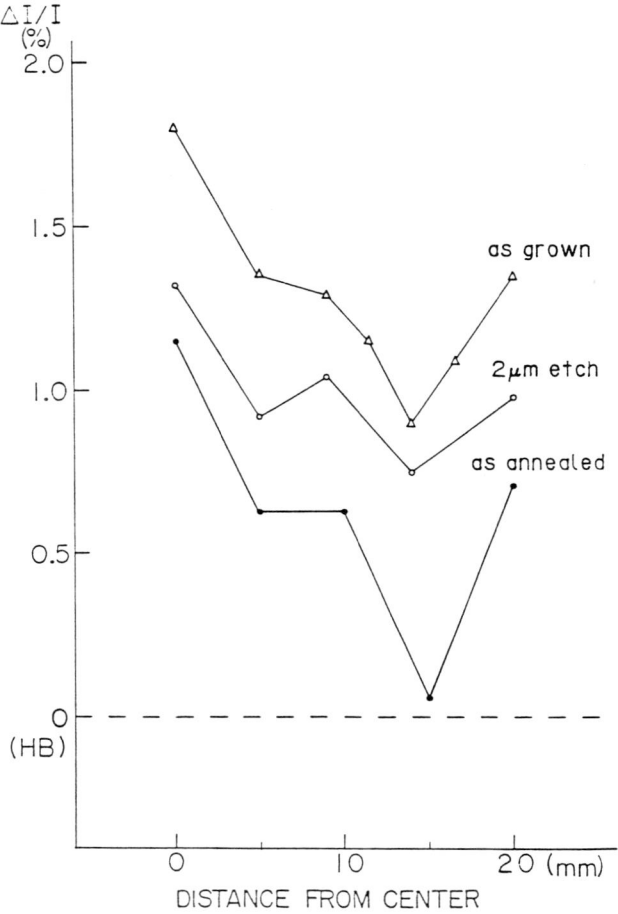

Fig. 8. Effect of heat treatment on the intensities of 200 reflection of undoped LEC (100) wafer. The value after etching by 4 μm coincides with that of as grown crystals. The results are shown for only half of the wafer.

PL intensity decreases. PL spectra at 77° K were found to be different for the two case $\gamma>2$ and $\gamma<1$, which implies some different origin for the similar behaviour of nonstoichiometry observed for the two γ region.

5. Conclusion

A new method of direct detection of nonstoichiometry and lattice location of doped impurities in GaAs was developed and its usefulness for the

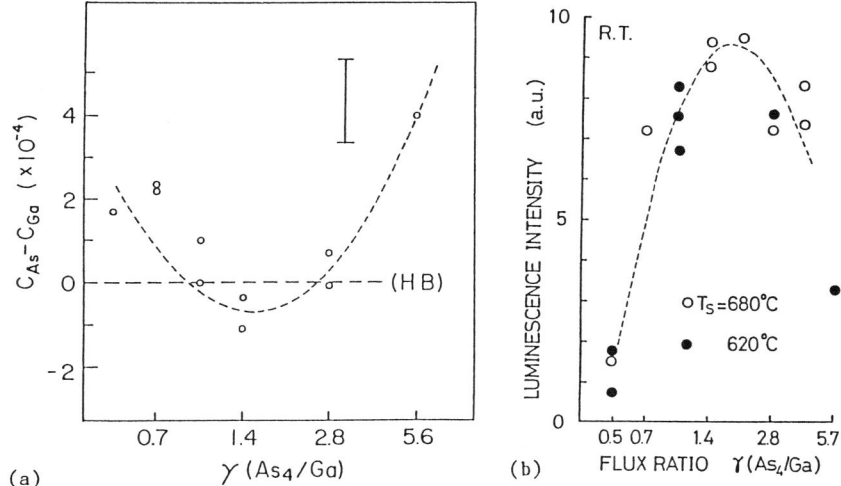

Fig. 9. Variations of nonstoichiometry (a) and PL intensity (b) of MBE grown layer with flux ratio γ.

characterisation of GaAs crystals grown by various methods was shown, though some problems are yet to be solved; identification of species of defects and determination of their microscopic structures. Usage of the X-ray wavelength around the absorption edge from synchrotron radiation would be one solution to those problems. For the detection of the defects with lower concentrations, an improvement of accuracy is necessary; the smaller effects such as diffuse scattering and lattice distortion are no longer negligible and should be taken into account. Further refined measurements with higher accuracy will provide us with an important knowledge on GaAs such as conversion of surface layers caused by ion implantation or by deposition of insulating films.

We can also apply the present method to other kinds of crystals such as InSb, ZnSe, CdTe, AlP and BN. Although its application to GaP and InP is difficult as can be seen from the principle of the present method, ternary and quarternary compounds like InGaAs, InGaAsP, CdHgTe, —, etc. with appropriate composition ratio can be investigated on such a subject as clustering.

The author wishes to thank Dr. K. Tada of Sumitomo Electric Industries, Ltd. and T. Takenaka of Shin-etsu Handotai Co., Ltd. for supplying crystals. Thanks are also due to Dr. H. Nakashima and Dr. T. Narusawa of Optoelectronics Joint Research Laboratory for the helpful discussion on the lattice location of doped Si and Zn. He also wishes to thank Dr. T. Suzuki, K. Kobayashi and N. Kamata for MBE growth, and Dr. J. Chikawa for his encouragement through this work.

REFERENCES

1) J. Nishizawa, Y. Okuno, and H. Tadano: J. Cryst. Growth **31** (1975) 215.
2) J. Parsey, Y. Nanishi, J. Lagowski, and H. Gatos: J. Electrochem. Soc. **129** (1982) 388.
3) D. Holmes, R. Chen, K. Elliot, and C. Kirkpatric: Appl. Phys. Lett. **40** (1982) 46.
4) L. Ta, H. Hobgood, A. Rohatgi, and R. Thomas: J. Appl. Phys. **53** (1982) 5771.
5) R. Logan and D. Hurle: J. Phys. Chem. Solids **32** (1971) 1739.
6) M. Straumanis and C. Kim: Acta Cryst. **19** (1965) 256.
7) H. Potts and G. Pearson: J. Appl. Phys. **37** (1966) 2098.
8) H. Kudo, Y. Ochiai, K. Takita, K. Masuda, and S. Seki: J. Appl. Phys. **50** (1979) 5039.
9) I. Fujimoto: Jpn. J. Appl. Phys. **23** (1984) L287.
10) I. Fujimoto: *Proc. 13th Int. Conf. on Defects in Semiconductors, Coronado*, ed. L. C. Kimerling and J. M. Parsey, (Metallurgical Society of AIME, 1984) p. 943.
11) H. Cole and N. Stemple: J. Appl. Phys. **33** (1962) 2227.
12) H. Nakashima, T. Narusawa, S. Semura, and K. Kobayashi: *Proc. 13th Int. Conf. on Defects in Semiconductors, Coronado*, ed. L. C. Kimerling and J. M. Parsey, (Metallurgical Society of AIME, 1984) p. 463.
13) J. Neave, P. Dobson, J. Harris, P. Dawson, and B. A. Joyce: Appl. Phys. **A32** (1983) 195.
14) T. Narusawa, Y. Uchida, K. Kobayashi, T. Ohta, M. Nakajima, and H. Nakashima: *Galliun Aarsenide and Related Compounds*, (Biarritz, 1984) Inst. Phys. Conf. Ser. No. 74, p. 127.
15) W. Spitzer and M. B. Panish: J. Appl. Phys. **40** (1969) 4200.
16) M. Greiner and J. Gibbons: Appl. Phys. Lett. **44** (1984) 750.
17) M. Tajima: Jpn. J. Appl. Phys. **21** (1982) L227.
18) S. Miyazawa, Y. Ishii, S. Ishida, and Y. Nanishi: Appl. Phys. Lett. **43** (1983) 853.
19) S. Makram-Ebeid, D. Gautard, P. Devillard, and G. M. Martin: Appl. Phys. Lett. **40** (1982) 161.
20) M. Heilbum, E. Mendez, and L. Osterling: J. Appl. Phys. **54** (1983) 6982.

Defects and Properties of Semiconductors: Defect Engineering, edited by J. Chikawa, K. Sumino, and K. Wada, pp. 87–98.
© KTK Scientific Publishers, Tokyo, 1987.

GROWTH OF DISLOCATION-FREE InP SINGLE CRYSTALS

Seiji SHINOYAMA

NTT Ibaraki Electrical Communication Laboratories, Tokai-Mura, Naka-Gun, Ibaraki-ken, 319-11, Japan

Abstract. This paper presents dislocation-free InP crystal growth by the liquid encapsulated Czochralski method. Nominally undoped dislocation-free crystals having a diameter of 15 mm are confirmed to be successfully grown under the low temperature gradient of 55°C/cm using a necking technique. In addition, the experimental relationship between the maximum diameter of the crystal obtained with a dislocation-free structure and the temperature gradient near the InP melt surface is shown. Finally, a technique combining the low temperature gradient with impurity hardening is demonstrated as a growth method for large-diameter dislocation-free crystals.

1. Introduction

Indium phosphide (InP) bulk crystals are becoming increasingly important as substrate materials for laser diodes[1,2] avalanche photodiodes[3] and for use in integrated opto-electronics.[4] As substrates, single crystals of InP are currently prepared by the liquid encapsulated Czochralski (LEC) method. These crystals generally contain a high density dislocations, or approximately $10^5/cm^2$. Recently, it has been reported that dislocations in substrates adversely affect the characteristics of devices.[5] Therefore, the development of a growth technique for dislocation-free crystals is strongly desired. Relative to this, this report describes the dislocation-free crystal growth of InP using the LEC method.

2. Problems in Growth of Dislocation-Free Crystals

The first problem concerning growth of dislocation-free InP crystals arises from the high equilibrium pressure of phosphorus (approximately 27.5 Kg/cm^{2} [6]) for InP at its melting point. This necessitates the use of the LEC technique under a high ambient gas pressure condition as the crystal growth method. The high gas pressure acts to increase the convection current of the gas in the growth chamber. This results in an abrupt temperature gradient occurring near the InP melt surface, which causes large thermal stresses in the growing crystal. Temperature profiles along the center axis of a crucible are schematically shown as a function of the gas pressure in Fig. 1.[7] As can be seen, the temperature gradient in the encapsulant (B_2O_3) layer increases with the N_2 gas pressure.

The second problem with respect to growth is the result of the small critical shear stress of InP. Table 1 presents lower yield stresses for the III–V materials measured at $0.65 T_M$ (T_M: melting point) by Gottschark et al.[8] It should be noted that the value of InP is very small in the III–V materials.

As InP is known to be a crystal that is easily twinned during growth, this twinning represents the third problem. The susceptibility to twinning depends on the stacking fault energy of the substance. The stacking fault energies of the III–V materials obtained by Gottschark et al.[8] are presented in Table 2. As indicated, InP has the smallest value in the III–V materials.

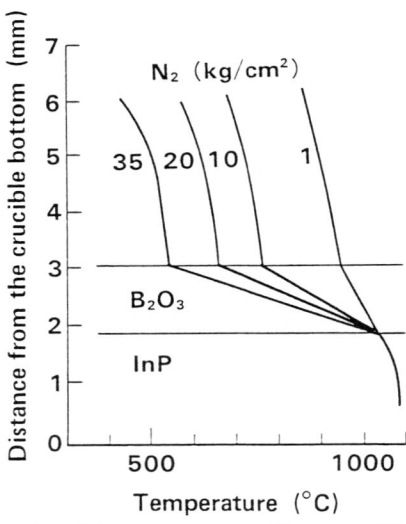

Fig. 1. Schematic illustration of the temperature profile in a crucible as a function of ambient gas pressure.

Table 1. Lower yield stresses for the III–V materials at $0.65 T_M$ (T_M: melting point). (N mm^{-2}).

GaSb	InSb	GaP	GaAs	InP	InAs
1.58	5.0	4.0	1.9	1.8	0.8

From Gottschark et al.[8]

Table 2. Stacking fault energies for the III–V materials. (meV/atom).

GaSb	GaAs	InSb	GaP	InAs	InP
53	47	43	33	30	17

From Gottschark et al.[8]

Twin formation occurs mostly in the growth of the shoulder part of the crystal. In this case, the twin (111) planes exhibit the configuration indicated in Fig. 2. Therefore, twinning can be avoided by ensuring that the enlargement angles of the shoulder are formed less than 19.5° and 35.3° for the <111> and <100> growth directions, respectively.[9]

Examples of such shape-controlled twin-free crystals are presented in Fig. 3. Although twins rarely appeared in such shape-controlled crystals, their origin when they did occur was confirmed to be thermal decomposition following growth. The twinning process itself is believed to involve the following steps. When a crystal is pulled outside the B_2O_3 layer, thermal decomposition begins at the crystal surface. Thermal decomposition rarely penetrates deeply into the interior of the crystal, but rather forms small decomposed cavities. Those cavities which reach the growth interface are responsible for the twin formation. Supression of such thermal decomposition after growth has been achieved with a thin protective skin of B_2O_3 on the crystal withdrawn from the B_2O_3 layer. Formation of the skin depends on the

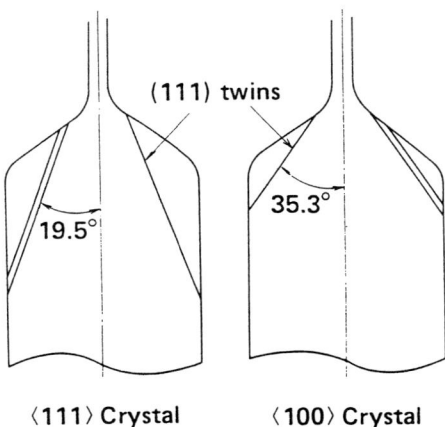

Fig. 2. Schematic illustration of twin boundaries appearing in the shoulder part of crystals.

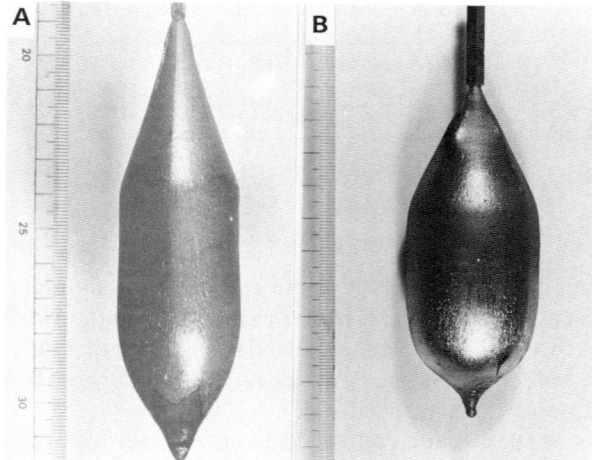

Fig. 3. Examples of twin-free crystals; cone angles are controlled within 19.5° and 35.3° for the <111> and <100> crystals, respectively.

viscosity of the B_2O_3 at the very surface of the B_2O_3 layer, which in turn depends on the temperature. It was found that the protective B_2O_3 skin was formed when its temperature at the layer's surface was less than 550°C. This temperature condition restricts the making of a low-temperature gradient near the InP melt surface which is needed for the reduction of thermal stress in a crystal during its growth.

3. Dislocation Densities and Distributions in Crystals

The origin of dislocation generation for semiconductor materials such as Si and Ge has been investigated, and it was pointed out that thermal stress was a main factor.[10] On the other hand, from an analysis of dislocation density distributions in GaAs crystals, Brice reported that aggregation of point defects is also a source of dislocation generation.[11] We studied the origin of dislocation generation in LEC-grown InP through an examination of dislocation densities and their distributions in crystals.

The relationship between dislocation density and crystal diameter is portrayed in Fig. 4. In this experiment, the crystals used for the samples were grown in the <111> directions under the same growth conditions simply varying the crystal diameter. As can be seen, the density is strongly dependent on crystal diameter, and increases in proportion with the 1.8th power of the diameter.

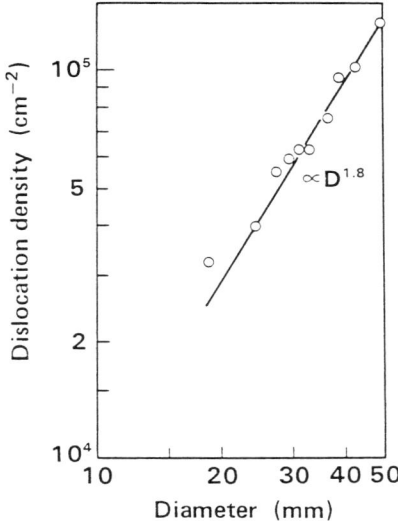

Fig. 4. Relationship between dislocation density and crystal diameter.

The radial distributions of dislocation density for several samples having various diameters, are presented in Fig. 5. All the samples reveal similar distributions, i.e., the density is highest at the edge, intermediate in the center and lowest in the middle region. The lowest positions, which are shown by arrows in the figure, are at about the halfway point of the radii for all the samples. Jordan et al.[12] carried out calculations of thermal stresses in <100> GaAs crystals during the LEC growth. From the results, they pointed out that thermal stress increased with crystal diameter and had a W-shaped profile in the radial direction.

The dependence of dislocation density on crystal diameter and the radial distribution of dislocation density obtained in this work seem to be similar to those of the thermal stress. To permit a detailed comparison of both, the thermal stress distribution in the <111> crystal was calculated. The calculation was performed based on the method reported by Jordan et al. Twelve resolved shear stresses for the (111), <110> slip system were obtained for the cylindrical <111> crystal, assuming that the temperature profile was parabolic in the radial direction and uniform along the crystal axis.

Figure 6 indicates the comparative result with respect to the <110> direction perpendicular to the <111> crystal axis. As can be seen, both curves coincide with each other. From this result, we concluded that dislocation density in the LEC InP is dominated by the thermal stresses existing in crystals.

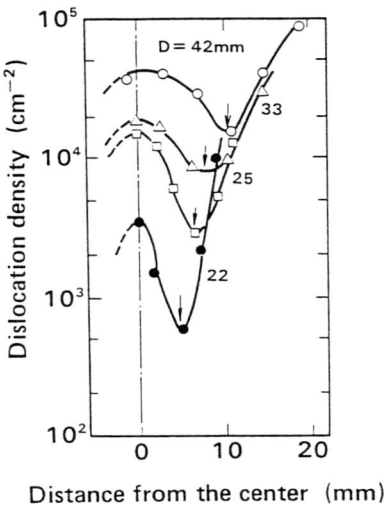

Fig. 5. Radial distributions of dislocation density along the <110> direction in (111) wafers; D in the figure indicates the crystal diameter.

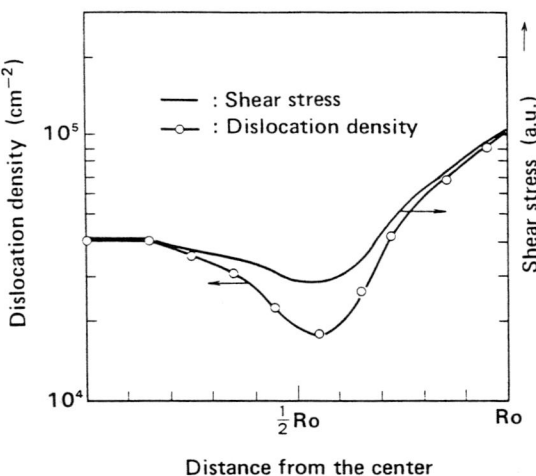

Fig. 6. A comparison between dislocation density and thermal stress along the <110> direction in a (111) wafer. Thermal stress shows the sum of the twelve shear stresses.

4. Growth of Dislocation-Free Crystals

Growth of nominally undoped dislocation-free crystals was performed by using a necking-in procedure. Necking conditions for the <111> growth direction were experimentally obtained to be a diameter of the neck part at less than 3 mm and a length longer than 5 mm. In order to decrease the thermal stresses in a crystal, it is essential to realize a low temperature gradient near the InP melt surface. As described previously, however, to avoid thermal decomposition after growth, the surface temperature of the B_2O_3 layer must be maintained at less than 550°C. Increasing the B_2O_3 thickness was found to be the most effective method to satisfy the above two contradictory conditions.[7] The temperature distributions between 6 mm-thick B_2O_3 and 24 mm-thick B_2O_3 are compared in Fig. 7.

Single crystals were grown under the five conditions presented in Table 3 with necking and the subsequent gradually increasing of the diameter. We examined the critical diameter, where dislocations were first generated after the necking. An example of an X-ray topograph of the part where dislocations are generated, is presented in Fig. 8. From the position marked with an A, dislocations are initially generated. The critical diameters obtained are presented in Table 3. The critical diameter increases with a decrease in temperature gradient. From the results, it is thought that dislocation generation occurs at the position where thermal stress becomes sufficient enough to cause dislocation gliding.

Dislocation-free crystals were successfully grown by limiting the crystal diameter to less than the critical diameter throughout the crystal length. An

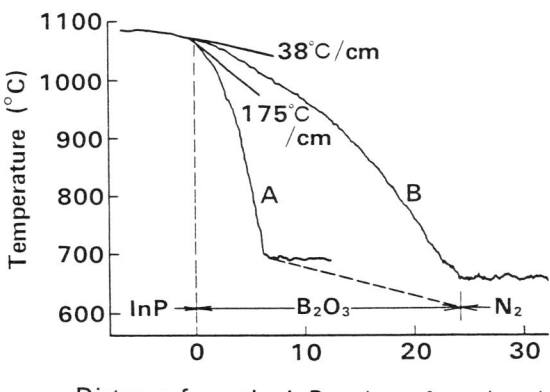

Fig. 7. Temperature profiles along the center axis in a crucible; B_2O_3 thickness is 6 mm (A) and 24 mm (B).

Table 3. Growth conditions and the corresponding critical diameters obtained from the experiments.

Condition	Thickness of B_2O_3	Temperature gradient	Critical diameter
1	6 mm	175° C/cm	8 mm
2	12	115	12
3	18	55	15
4	24	38	20
5	40	25	22

example of a dislocation-free crystal having a diameter of 15 mm and a length of 150 mm which is grown under the temperature gradient of 55°C/cm, is shown in Fig. 9. It was also confirmed that Sn-doped (n-type) and Fe-doped (semi-insulator) dislocation-free crystals could also be obtained under the same growth conditions.

5. Approach to Large Diameter Dislocation-Free Crystals

In the previous section, we demonstrated the possibility of growing dislocation-free crystals having a diameter of 15 mm. In actual substrate use, however, larger diameter crystals are desirable.

Figure 10 shows a plotting of the relationship between the critical diameter, D_c, and the temperature gradient, $(\partial T/\partial z)_0$, presented in Table 3. From the relationship, the following empirical formula was obtained:

$$(\partial T/\partial z)_0 \cdot D_c^2 = C \text{ (constant)},$$

where $C = 124$ deg·cm. The solid line is a calculation curve. Using the curve, temperature the gradient for a 50 mm diameter dislocation-free crystal is estimated to be about 5°C/cm. Under such a low temperature gradient, it is presumed that control of crystal growth will become very difficult.

On the other hand, it is known that the high concentration doping of specific impurities also makes it possible to obtain dislocation-free crystals. To date, such impurities as S, Zn, Te, Ga, Sb and As have been reported.[13,14] In order to enlarge the diameter of dislocation-free crystals, combining the two methods would be more effective, i.e., the impurity doping method and the low temperature gradient method described above. In the present work, S,

Fig. 8. An X-ray topograph of the region where dislocations are generated; from the part marked by A, dislocations are generated.

Ga and Sb were used as the doping impurities. Seki *et al.* reported that S, an n-type dopant, is an impurity considerably effective in reducing dislocation generation in doped concentrations above $5 \times 10^{18}/\text{cm}^3$.[13] On the other hand, Ga and Sb, which are isoelectric impurities, have been reported by Jacob[14] to have the same effects, although the details are not yet clear.

Fig. 9. Example of a nominally undoped dislocation-free crystal. The diameter is about 15 mm.

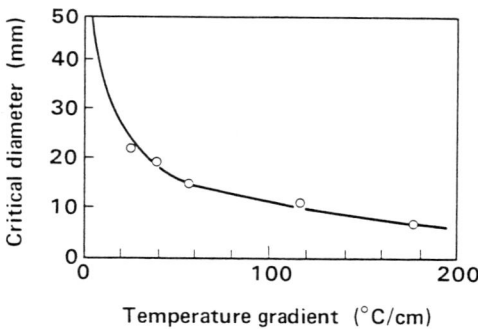

Fig. 10. Relationship between the temperature gradient at the InP melt surface and the critical diameter; the solid line is a calculation curve.

We therefore studied the concentration dependence on the effect and the segregation coefficients of Ga and Sb, and obtained two results. First, both impurities exhibited the dislocation reduction effects in concentrations above $5 \times 10^{18}/\text{cm}^3$. Second, the segregation coefficients were 3.35 and 0.12 for Ga and Sb, respectively.[15] From these results, particularly considering the latter, Ga and Sb co-doping is found to be more effective because the impurity content can be maintained with a high concentration (total concentration of Ga and Sb) throughout the crystal length.

Figure 11 show the radial distributions of the dislocation density for impurity doped crystals which are grown to a diameter of 40 mm under the temperature gradient of 55°C/cm. The result for an undoped crystal grown under the same growth condition is also presented in the figure for comparison. Dislocation densities of the doped crystals are remarkably low compared to that of the undoped crystal. In the doped crystals, dislocation-free regions are present in their centers. These regions were maintained throughout the crystal length having nearly constant diameters. The critical diameters were determined to be 30 mm and 20 mm for the S-doped and the Ga and Sb co-doped crystals respectively. These values are significantly higher than that for the undoped crystal ($D_c = 15$ mm).

6. Conclusion

Growth of dislocation-free InP crystals using the LEC method has been investigated. From the good agreement between dislocation density distribution and thermal stress distribution obtained by calculation, it was confirmed that dislocation densities in the LEC-grown InP crystals were dominated by thermal stresses. Increasing the B_2O_3 thickness was found to be the most

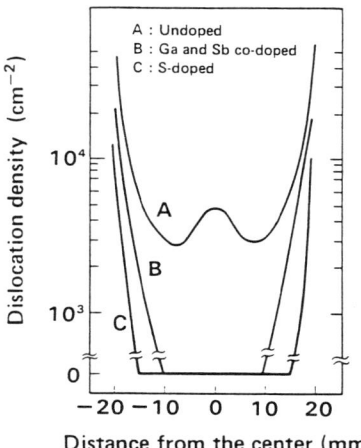

Fig. 11. Radial distributions of dislocation density; (A): undoped crystal; (B): Ga and Sb co-doped crystal; total concentration of Ga and Sb is $1 \times 10^{19}/cm^3$; (C): S-doped crystal; the concentration is $5 \times 10^{18}/cm^3$.

effective for decreasing the temperature gradient near the InP melt surface. As a result, nominally undoped dislocation-free crystals having a 15 mm diameter were successfully grown. In order to further enlarge the diameter of dislocation-free crystals, the growth technique combining the two method, i.e., the impurity doping method and the low temperature gradient method, was demonstrated.

REFERENCES

1) J. J. Hsieh, J. A. Rossi, and J. P. Donnelly: Appl. Phys. Lett. **30** (1977) 353.
2) H. Nagai, Y. Noguchi, K. Takahei, Y. Toyoshima, and G. Iwane: Jpn. J. Appl. Phys. **19** (1980) L218.
3) Y. Takanashi and Y. Horikoshi: Jpn. J. Appl. Phys. **18** (1979) 2173.
4) J. P. Noblanc: *Semi-Insulating III–V Materials, Evian, 1982* (Shiva Publishing Limited, Kent, 1982) p. 380.
5) N. Susa, Y. Yamaguchi, and H. Ando: J. Appl. Phys. **53** (1982) 7044.
6) K. J. Bachmann and E. Buehler: J. Electron. Mater. **3** (1974) 279.
7) S. Shinoyama, C. Uemura, A. Yamamoto, and S. Tohno: Jpn. J. Appl. Phys. **19** (1980) L331.
8) H. Gottschark, G. Patzer, and H. Alexander: Phys. Status Solidi. (a) **45** (1978) 207.
9) S. Shinoyama: *IOOC'83 Post Conference Meeting Technical Digest* (1983) 9.
10) E. Billing: Proc. Roy. Soc. A. **235** (1956) 37.
11) J. C. Brice: J. Cryst. Growth **7** (1970) 9.
12) A. S. Jordan, R. Caruso, and A. R. Von Neida: The Bell Syst. Tech. J. (1980) 593.
13) Y. Seki, J. Matsui, and H. Watanabe: J. Appl. Phys. **47** (1976) 3374.
14) G. Jacob: *Semi-Insulating III–V Materials, Evian, 1982* (Shiva Publishing Limited, Kent, 1982) p. 2.
15) S. Tohno, E. Kubota, S. Shinoyama, A. Katsui, and C. Uemura: Jpn. J. Appl. Phys. **23** (1984) L72.

InP MISFETS TECHNOLOGY

T. SUGANO

Department of Electronic Engineering, The University of Tokyo, 7-3-1 Hongo, Bunkyo-ku, Tokyo, 113, Japan

Abstract. InP metal-insulator-semiconductor field-effect transistors were fabricated using plasma anodic aluminum oxide as the gate insulator. The effective electron mobility in the surface channel is 2600 cm^2/V·s at room temperature. A good stability in the drain current-voltage characteristic was obtained using the plasma anodic aluminum oxide and interlayed native oxide as the gate insulator.

1. Introduction

Since the feasibility of surface accumulation type InP MISFETs was demonstrated using pyrolytically deposited SiO$_2$ film as the gate insulator,[1] InP MISFETs with various kinds of deposited insulating film such as Al$_2$O$_3$ film have been reported.[2]

Also the possibility of surface oxidizing InP wafers thermally[3] or anodically in aqueous solution[4] was explored. So far plasma anodization technique has been proved to be promising dry growth technology of the oxidation on Si[5] or GaAs[6] substrates at a low temperature, but oxidation rate was found to be very small for InP. On the other hand, it was found that anodization of evaporated Al-GaAs systems gave MIS diodes with good interface properties,[7,8] so that evaporated Al-InP systems have been oxidized anodically in oxygen plasma by the author's group.

Here, technology of anodization of evaporated Al-InP system in oxygen plasma, physical properties of the aluminium oxide film, in-depth profile of the film composition by Auger Electron Spectroscopy (AES), trap state density at the interface between aluminium oxide and InP, static characteristics of experimental aluminium oxide and InP-MISFETs, and also improvement of drift in the drain current-voltage characteristics will be reported.

2. Sample Preparation

InP wafers whose surface orientation was (100) were used. As a first step of processing, the wafers were etched in Br-methanol mixture for a few minutes and rinsed in methanol and de-ionized wafer, successively. Aluminium was evaporated by a tungsten heater in a conventional vacuum station with a diffusion-pump at the pressure of $\sim 8 \times 10^{-6}$ Torr, and deposited on InP wafers at almost room temperature. The anodization rate and properties of aluminium oxide film were very sensitive to the deposition rate of aluminium. In consequence, the deposition rate was controlled to 5 Å/sec and the thickness of aluminium film used in this experiment was 250 Å. The plasma anodization equipment which was used for anodization of aluminium film is the same as used for plasma anodization of GaAs and Si.[9] The plasma anodization was carried out at the constant anodization current density of ~ 6 mA/cm^2 in the oxygen of 0.1 Torr in pressure. The oxygen gas was moistened by passing it through a water bubbler whose temperature was kept at 40°C. The substrate temperature during anodization was between 150°C and 180°C. The terminal voltage between the anode and the cathode increased in the initial anodization period and saturates with time when the anodization of aluminium film was finished, as shown in Fig. 1, because the growth rate of the native oxide of InP is extremely small. The anodization has been stopped a few minutes after the terminal voltage reaked to plateau. If the anodization is

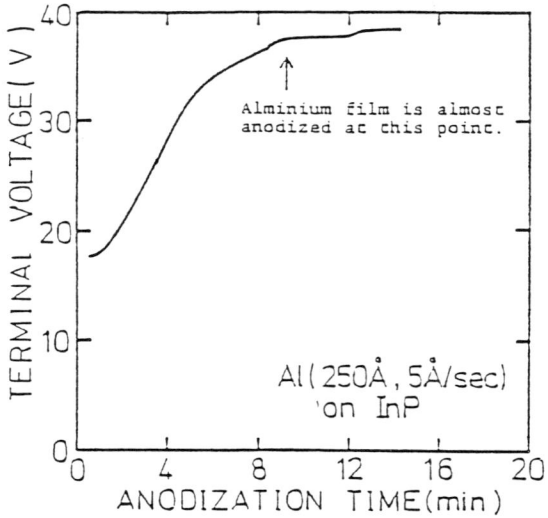

Fig. 1. The change of the terminal voltage during plasma anodic oxidation. The terminal voltage is the dc voltage between the anode and the cathode when constant anodization current is following. The increase of the terminal voltage saturates after finishing anodization of Al on the surface of InP. Reproduced with permission of The Metallugical Society of AIME, © 1982 AIME. After Y. Hirayama, H. M. Park, F. Koshiga and T. Sugano, Journal of Electronic Materials; Vol. 11, No. 6, 1982, p. 1011–1022.

continued for a long time after the saturation of terminal voltage, the resistivity of the oxide film becomes small. After anodization, annealing was carried out in argon at 250°C for 30~90 minutes.

3. Properies of Oxide Film and Interface

3.1 Properties of plasma anodic oxide film

The measurements of dielectric properties of plasma-anodic aluminium oxide film were carried out using MIS diodes fabricated on n-InP substrates, whose carrier concentration was 10^{18} cm^{-3}. The typical physical properties of the oxide film are shown in Table 1. The thickness and the refractive index were measured by ellipsometry, assuming that the transparent oxide film on InP is homogeneous in thickness and optical properties and that InP substrate has the complex refractive index of 3.42–0.38 j. The dielectric constant was evaluated from capacitance per unit area of MIS diode biased to strong accumulation.

The result of the AES analysis of a sample are shown in Fig. 2. Phosphorus is incorporated into the oxide film, but the incorporation of indium into the oxide film is small. Carbon, which is detected on the top surface of the oxide, is due to contarmination. When the anodization has continued after the terminal voltage has reaked to plateau, indium is more incorporated into the oxide film. Auger chemical shift shows that indium

Table 1. Typical properties of plasma anodic Al$_2$O$_3$ on InP substrate.

	Time of anodization after saturation	
	2–3 min	12–15 min
Thickness	630 Å	860 Å
Refractive index (6328 Å)	1.72	1.68
Resistivity (2×10^{-6} A/cm^2)	5×10^{10}~10^{12} Ω·cm	3×10^{9}~2×10^{10} Ω·cm
Dielectric constan	~7	~6
Breakdown strength	1.2~1.6$\times 10^6$ V/cm	0.8~1$\times 10^6$ V/cm

The thickness of the aluminium film was about 250 Å and deposition rate of aluminium was 5 Å/sec. Reproduced with permission of The Metallurgical Society of AIME, © 1982 AIME. After Y. Hirayama, H. M. Park, F. Koshiga and T. Sugano, Jounal of Electronic Materials, Vol. 11, No. 6, 1982 p. 1011–1022.

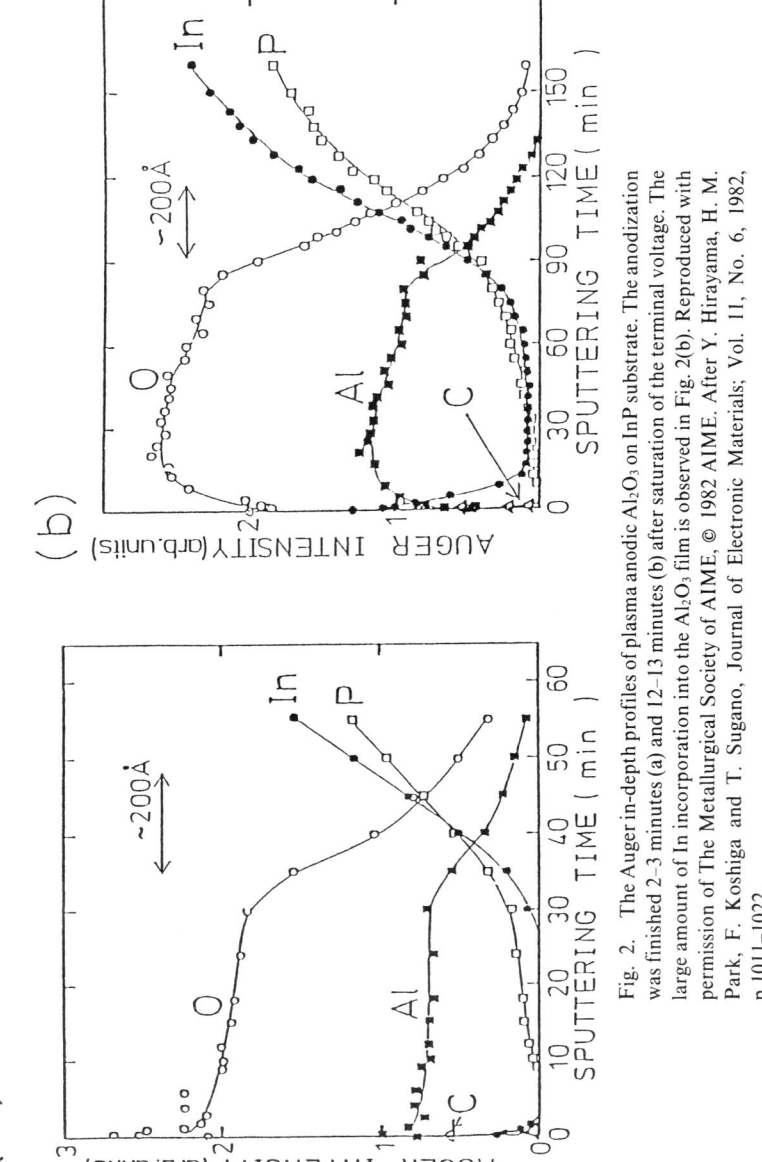

Fig. 2. The Auger in-depth profiles of plasma anodic Al$_2$O$_3$ on InP substrate. The anodization was finished 2–3 minutes (a) and 12–13 minutes (b) after saturation of the terminal voltage. The large amount of In incorporation into the Al$_2$O$_3$ film is observed in Fig. 2(b). Reproduced with permission of The Metallurgical Society of AIME, © 1982 AIME. After Y. Hirayama, H. M. Park, F. Koshiga and T. Sugano, Journal of Electronic Materials; Vol. 11, No. 6, 1982, p.1011–1022.

atoms in the oxide film is not oxidized, while indium atoms near the top surface of the oxide are oxidized. The oxide film, which incorporated elemental indium atoms, has poor resistivity and breakdown strength. This suggests that elemental indium atoms incorporated into oxide film are responsible for poor electrical properties of the oxide film.

3.2 Capacitance-voltage characteristics

MIS diodes used in C-V measurement were fabricated on n-InP substrates whose carrier concentration was $3.8 \sim 4 \times 10^{15}$ cm^{-3} and p-InP substrates whose carrier concentration was $2 \sim 2.3 \times 10^{16}$ cm^{-3}. To make backside ohmic contact, Au-Ge/Ni/Au was deposited for n-InP substrate. The final annealing process at 250°C after the plasma anodization was effective to make the contact ohmic. For p-InP substrate, Zn/Au was deposited and then annealed at 350°C for 5 minutes in argon ambient. Figure 3 is the C-V characteristics of an aluminium oxide p-InP MIS diode measured in dark. The theoretically calculated minimum value of capacitance is also shown.

Assuming that 10MHz is a sufficiently high frequency to measure the C-V characteristics, the distribution of the density of the interface trap states in the band gap of InP has been obtained, and shown in Fig. 4. In the same figures, closed circles indicate the density of interface trap state calculated by assuming that the flat band voltage shift with temperature is due to the interface states.[10] The density of states is smaller than 10^{12} cm^{-2} eV^{-1} over the broad range near the midgap and the minimum density of states is 4×10^{11} cm^{-2} eV^{-1}.

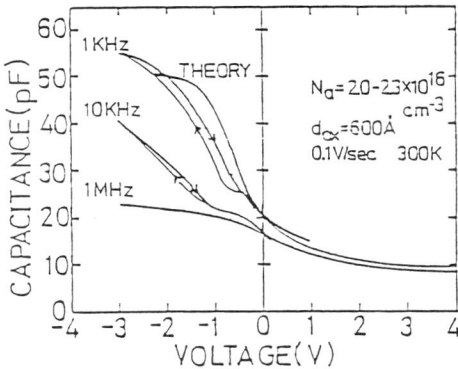

Fig. 3. The C-V characteristics of plasma anodic aluminium oxide p-type InP MIS diode measured in dark at 300 K. Reproduced with permission of Institute of Physics, © 1981. After Y. Hirayama, H. M. Park, F. Koshiga and T. Sugano, Institute of Physics, Conference Series No. 13, p. 341–346.

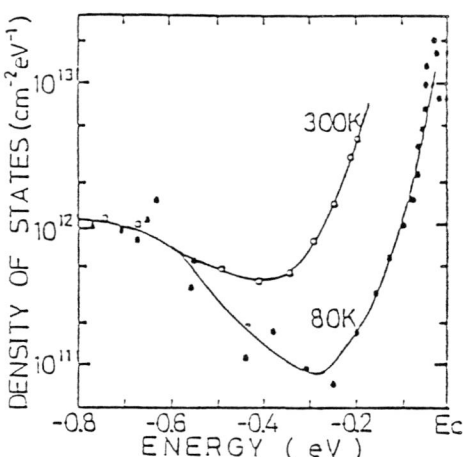

Fig. 4. The distribution of the density of the interface trap states in the band gap of InP.

3.3 DLTS measurement

Here, the DLTS data from plasma anodic aluminium oxide n-type InP MIS diode are presented.

Figure 5 shows the DLTS spectrum measured by maintaining Va and Vb at constant value through temperature scan. The DLTS signals for various Vb

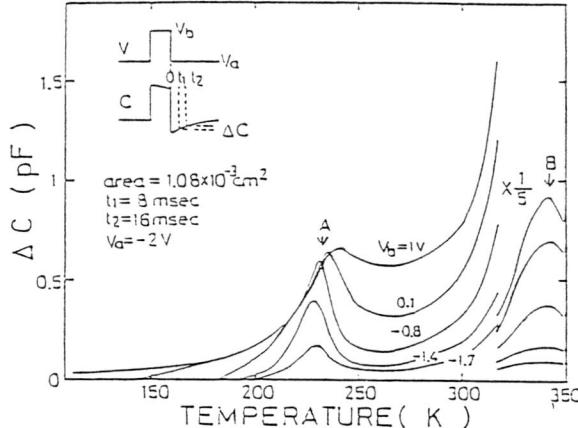

Fig. 5. DLTS data measured by the constant voltage mode. Reproduce with permission of Institute of Physics, © 1981. After Y. Hirayama, H. M. Park, F. Koshiga and T. Sugano, Institute of Physics, Conference No. 13, p. 341–346.

and sampling time were stored into memory during one temperature scan. Because the flat-band voltage of MIS diode becomes large with decreasing temperature, the InP surface is depleted enough below room temperature but it is not at higher temperature when Va is set to -2 volts. In Fig. 5 the contribution of bulk traps to DLTS signals is significant in comparsion with that of the interface states below room temperature because the InP surface is completely depleted.[11] The two peaks are shown in Fig. 5. The peak named A is assumed to be bulk electron trap which has the energy level of $0.45 \sim 0.55$ eV measured from the conduction band and the capture cross section of 5×10^{-16} cm^2, obtained from Arrhenius plot lne_n/T^2 vs $1/T$, where e_n is the emission rate of electron and T is the temperature. The density of traps is estimated $1 \sim 3 \times 10^{14}$ cm^{-3} by assuming homogeneous distribution of traps in depth. The peak named B appears at the same temperature as the one of deposited SiO$_2$-InP MIS diodes,[12] which have the energy level of 0.6–1.0 eV below the conduction band. In case of the peak named B, Arrhenius plot is inaccurate because of charging up of traps in the oxide at high temperature and the small peak shift by changing sampling times of t_1 and t_2. Below 200 K the DLTS signal is not observed at small Vb, at which the Fermi-level at the surface does not reach to the mid-gap. Therefore, the DLTS signal in low temperature range is due to the emission from the interface states.

4. InP MIS Diodes with Plasma Anodic Al$_2$O$_3$ and Interlayed Native Oxide

In order to improve the stability of InP MISFETs, two apparently contradicting approaches have been proposed. The first approach is to remove a native oxide film from the InP substrate surface by a vapor etching before depositing film.[13] However, such surface treatments may degrade the surface properties of the InP substrates, and thus the characteristics of MISFET's. The other approach is to grow a native oxide film between anodic Al$_2$O$_3$ and the InP substrate.[14] The latter focuses on growing high-quality native oxide film without slow trapping centers at the interface. The slow trapping of electrons at the interface between the native oxide and the thick overlaying insulator is also reduced by growing the native oxide film which is thick enough to prevent tunneling.

InP MISFET's were fabricated by growing a thin native oxide film with plasma anodization before depositing an aluminium film, which is to be anodically oxidized in order to make the overlaying gate insulator.

The fabrication steps of the MIS diodes areas are as follows. Bromine-methanol eched (100)-oriented wafers of undoped n-type ($\sim 9 \times 10^{15}$ cm^{-3}) InP were used as the starting material. A native oxide layer is grown on the InP substrate by plasma anodic oxidation. Under the anodization conditions used here, the thickness of the native oxide saturated nearly at 10 nm. Aluminium film of 25-nm in thickness was evaporated in a conventional vacuum station.

The deposition rate was controlled to 0.5 nm/s, because the density of the interface trap states depends on the deposition rate as shown in Fig. 6. Then, plasma anodization of Al film was carried out. The condition of anodic growth of the native oxide film is the same as in the plasma anodization of Al, except that the anodization must be finished a few minutes earlier than in the case of anodization of Al. Effect of annealing temperature on the density of interface trap states is illustrated in Fig. 6. Furthermore, the distribution the density of interface trap states of InP MIS diodes with different oxide are shown in Fig. 7.

The processing steps are shown in Fig. 9.

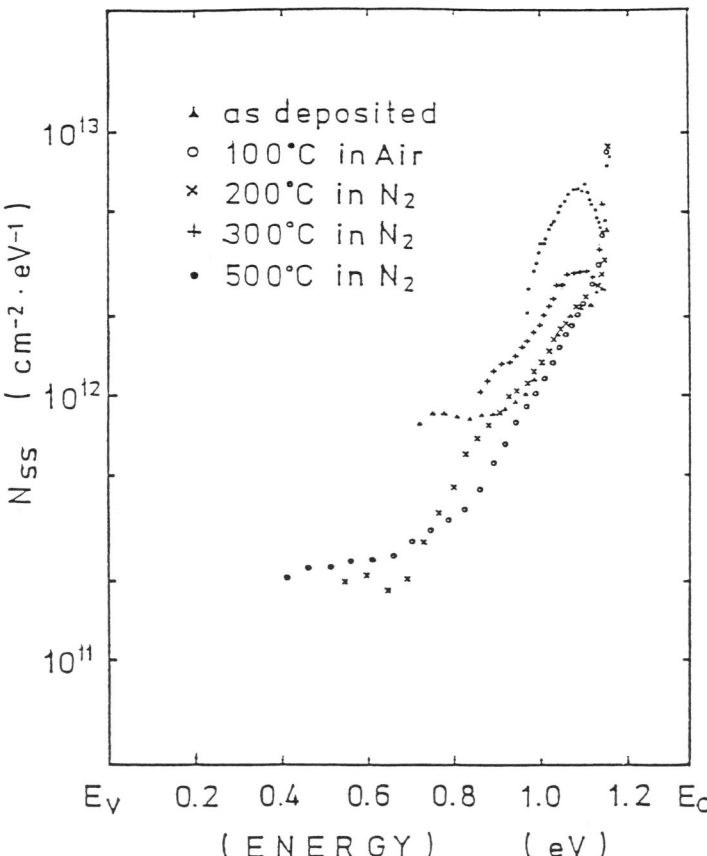

Fig. 6. Effect annealing temperature on the density of trap states of InP MIS diodes. The optical annealing temperature is rounged from 200°C to 300°C.

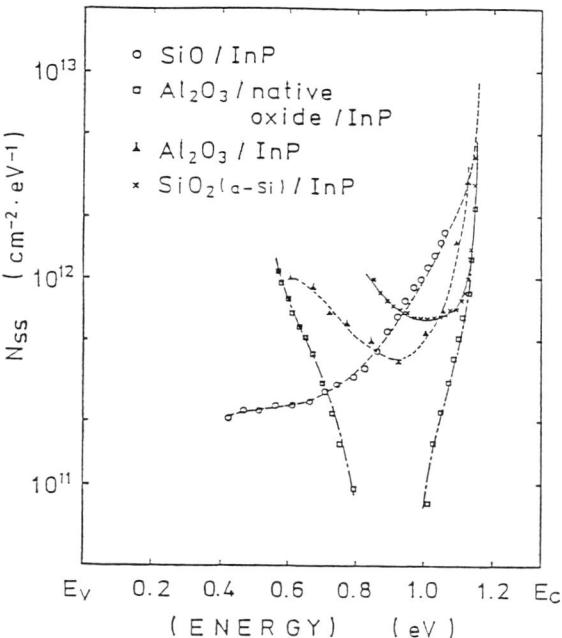

Fig. 7. The distributions of the density of interface trap states of InP MIS diodes with different oxide. The InP MIS diode with Al_2O_3 and interlayed native oxide has the smallest density of trap states.

Silicon ions whose dose was $5 \times 10^{13}/cm^2$ were implanted into the substrate at 100 keV. The substrate was heated to about 100°C during implantation. This ion implantation at elevated temperature is necessary for obtaining good quality n-type layers. Annealing of the implanted layer was carried out in forming gas ambient (N_2: 95 percent, H_2: 5 percent) at 700°C for 15 min. Phosphosilicate glass (PSG) film of 500-nm thickness was deposited on both sides of the wafer by plasma-assisted chemical vapor deposition (P-CVD) as a protective layer during activation. After removing the n layer from the gate region, in order to reveal the semi-insulating InP, the InP substrate was anodically oxidized in oxygen plasma to grow the 10-nm native oxide film. Then aluminium was deposited on the InP substrate and anodically oxidized, and the annealing was carried out at 350°C. This annealing temperature was chosen as a compromise between the hysteresis in the C-V characteristics and the density of the interface state. The channel length is 40 μm, the channel width is 100 μm, and the oxide capacitance is 80 nF/cm.

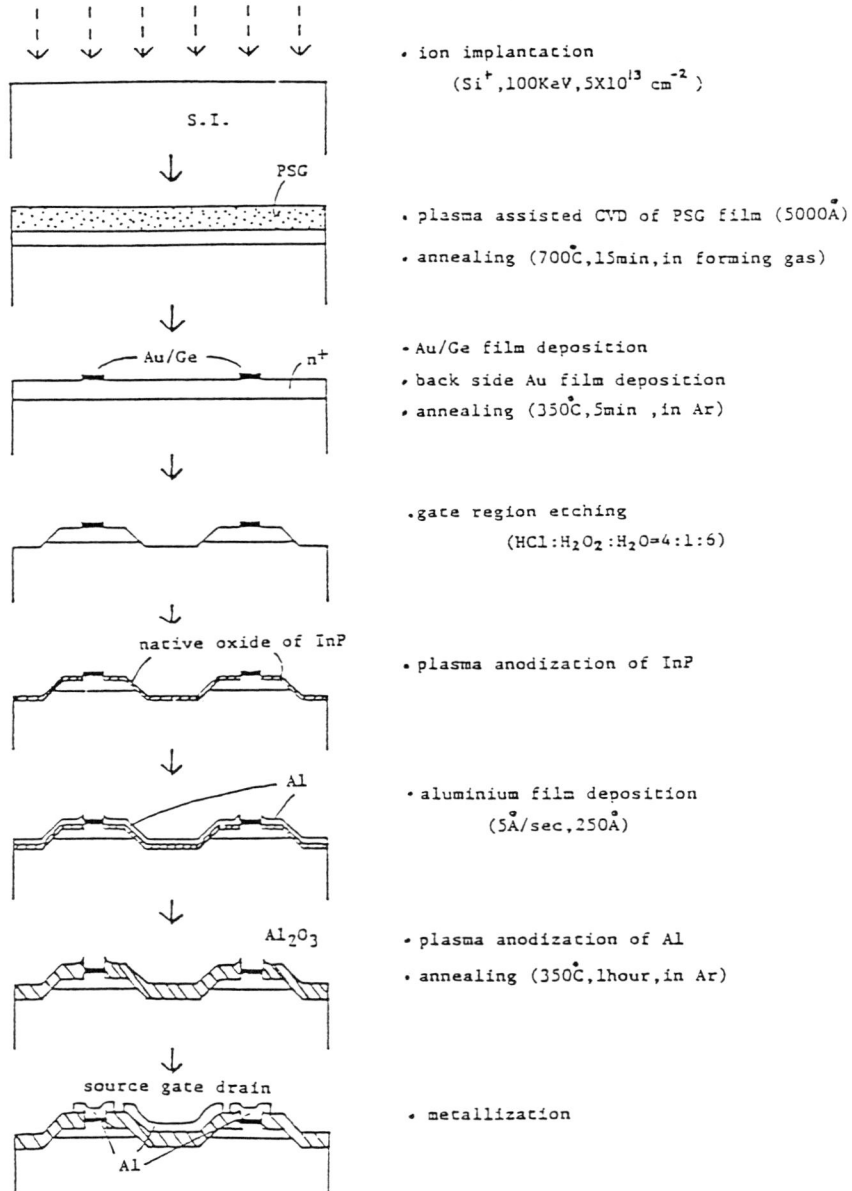

Fig. 8. Processing step of InP MISFETs fabrication.

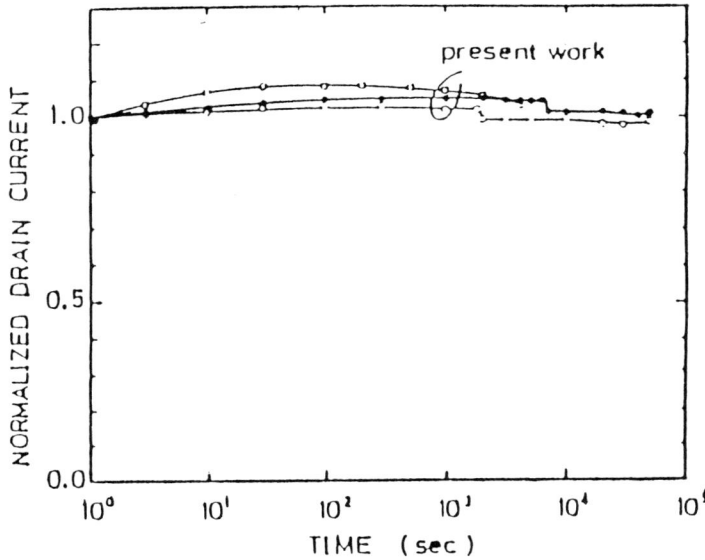

Fig. 9. Long term stability of the drain current of InP MISFET's which have native oxide film interlayed between plasma anodic Al_2O_3 film and InP substrate.

The output characteristics of MISFET are shown in Fig. 4. MISFET's are normally-on and leakage current is observed between the source and the drain, because the flat-band voltage is negative and the whole surface of the substrate shows n-type conduction. The effective electron mobility μ_{eff} and the threshold voltage V_{th} are found to be $2100 \sim 2600$ $cm^2/V \cdot s$ and $-0.3 \sim -0.9$ V at 300 K, respectively.

The long-term drifting behavior (1 s $\sim$ 15 h) of some devices is shown in Fig. 9. The current without leakage current is normalized to the same response at a time of 1 s. The applied drain voltage and gate voltage are 2 and 1.5 V, respectively. Some devices on another chip show a slight increasing-type drift, but the long-term stability of these MISFET's is small in comparison with that of the MISFET's using plasma anodic Al_2O_3 without the interlayed native oxide film.[15]

6. Conclusion

InP MISFET's with very little drifting of the drain current at room temperature have been fabricated using the gate insulator of plasma anodic Al_2O_3 film with an interlayed native oxide film. Effective electron mobility was 2600 $cm^2/V \cdot s$ and the variation of drain current with period of 5 μs to 5×10^4 s within ± 4 percent were obtained.

As compared with previously published results, the present MISFET's show significantly reduced drift and hopefully will have a positive impact on the fabrication technology of InP MISFET's.

Acknowledgment

The author would like to thank his colleagues, Dr. F. Arai, Dr. Y. Hirayama, Dr. H. M. Park and Mr. M. Matsui for their collaborations.

REFERENCES

1) D. L. Lile, D. A. Collins, L. G. Meiners, and L. Messick: Electron. Lett. **14** (1978) 657.
2) T. Kawakami and M. Okamura: Electron. Lett. **15** (1979) 502; D. V. Lang: J. Appl. Phys. **45** (1974) 3023.
3) J. F. Wagnar and C. W. Wilmsen: J. Appl. Phys. **51** (1980) 812.
4) D. L. Lile and D. A. Collins: Appl. Phys. Lett. **28** (1976) 554.
5) V. Q. Ho and T. Sugano: IEEE Trans. Electron Dev. **ED-28** (1981) 1060.
6) T. Sugano, F. Koshiga, K. Yamasaki, and S. Takahashi: IEEE Trans. Electron Dev. **ED-27** (1980) 449.
7) D. W. Langer, F. L. Schuermeyer, R. L. Johnson, H. P. Singh, C. W. Litton, and H. L. Hartnagel: J. Vac. Sci. Technol. **17** (1980) 964.
8) Y. Hirayama, F. Koshiga, and T. Sugano: J. Appl. Phys. (1981).
9) K. Yamasaki and T. Sugano: Japan J. Appl. Phys. (1979) Suppl. 17-1 321.
10) P. V. Gray and D. M. Brown: Appl. Phys. Lett. **15** (1966) 31.
11) K. L. Wang: IEEE Trans. Electron Dev. **ED-27** (1980) 2231.
12) J. Stannard: J. Vac. Sci. Technol. **16** (1979) 1462.
13) M. Okamura and T. Kobayashi: Japan J. Appl. Phys. **19** (1980) 2143.
14) T. Sawada and H. Hasegawa: Electron. Lett. **18** (1982) 742.
15) M. Matsui, Y. Hirayama, F. Arai, and T. Sugano: IEEE Electro Devices Letters **EDL-4** (1983) 308.

CHARACTERIZATION OF ALLOY SEMICONDUCTORS

T. KATODA

Institute of Interdisciplinary Research, Faculty of Engineering, The University of Tokyo, 4-6-1 Komaba, Meguro-ku, Tokyo 153, Japan

Introduction

Disorder in atomic arrangement is one of the most complicated but basic problem accompanied to alloy semiconductors. Unfortunately we have had no means to characterize clusters in an alloy semiconductor. Internal stress is expected to be accumulated in each bond included in some alloy semiconductors. Generally speaking strength of bond is different for each bond. It is very important to have methods to characterize clusters, internal stress and bond strength in order to study defects in an alloy semiconductor. The value of elastic stress in $Ga_{1-x}In_xAs_yP_{1-y}$ lattice matched to an InP substrate measured by luminescence polarization has been reported by Bert et al.[1] Although behavior of long wavelength optical phonons[2] can be used to characterize a clustering parameter in ternary alloy, they depend also on stress accumulated in atomic bonds. In this paper, a use of laser Raman spectroscopy will be proposed to estimate internal stress accumulated in each bond and a clustering parameter at the same time. It is possible to make clear difference in strength and disoder among bonds in alloy semiconductors with laser Raman spectroscopy.

At first characterization of both a clustering parameter and internal stress for some kinds of III–V alloy semiconductors will be described. Characterization of difference in disorder and annealing behavior between Ga–As and In–As bonds in $Ga_{0.47}In_{0.53}As$ will be reported. The behavior is thought to be related closely to the strength of bonds.

Figure 1 shows the process for characterizing a clustering parameter and stress accumulated in each bond included in alloy semiconductors with laser Raman spectroscopy presented in this paper. A clustering parameter β can be estimated by analyzing temperature dependence of intensity ratio between optical phonons as described in detail in Section 3. Stress accumulated in each bond is obtained from temperature dependence of optical phonon frequency.

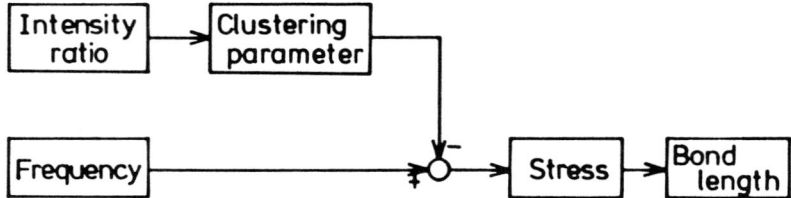

Fig. 1. Diagram of process to obtain clustering parameter β, stress and bond length.

A bond length corresponding to the stress is derived and is compared with that measured by Extended X-ray Absorption Fine Structure (EXAFS).

I. CHARACTERIZATION OF A CLUSTERING PARAMETER AND INTERNAL STRESS

1. Theoretical Calculation

1.1 Clustering parameter

The term of cluster in this paper means a lump of atoms of the same kinds caused by deviation from random distribution of atoms. It is assumed that all atoms are in their lattice sites, that is, lattices are maintained even when clusters are formed. Large clusters are due to microscopic phase separation and not to decomposition of an alloy such as spinodal decomposition. Clusters can be discussed quantitatively with a clustering parameter β which was introduced by Yamazaki et al.[2] because sizes and a density of clusters can not be measured directly at present. Verleuer et al.[3] reported the behavior of phonon in ternary alloy using REI model with the clustering parameter. By using the clustering parameter, for instance in the case of $Ga_{1-x}In_xAs$ alloys, the probability of finding gallium atom next to another gallium atom, P_{gg}, and that of finding indium atom, P_{ii}, are expressed as,

$$P_{gg} = 1 - x + x\beta$$
$$P_{ii} = x + (1 - x)\beta \quad (1)$$

P_{gg} and P_{ii} are equal to $1-x$ and x, respectively when $\beta=0$. When $\beta=1$, P_{gg} and P_{ii} are equal to one. At that time the atoms of the same species in cation site, that is gallium or indium in this case, are most likely to concentrate. The probability f_i of finding the i-th basic unit cell which constitute an alloy as shown in Fig. 2 can be calculated with a clustering parameter as follows. The probabilities are expressed as

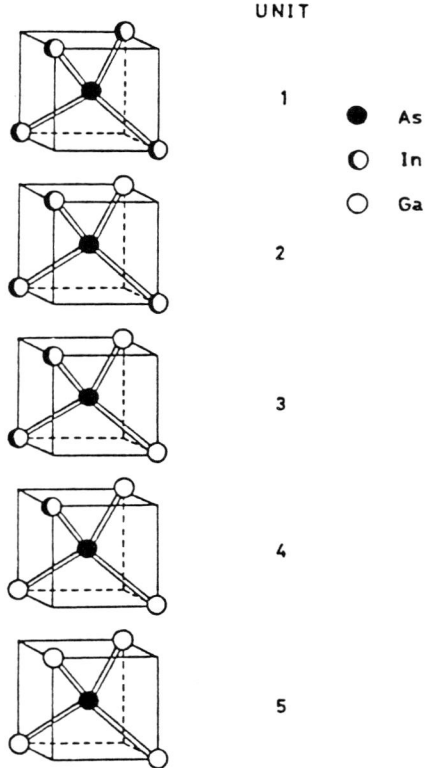

Fig. 2. Basic units of nearest-neighbor around an As atom.

$$f_1 = (1 - x)\{P_{gg} - 1 + x + (1 - x)P_{gg}^2\}$$

$$f_2 = 4(1 - x)^2 P_{gg}(1 - P_{gg})$$

$$f_3 = 6x^2 P_{ii}(1 - P_{ii}) \qquad (2)$$

$$f_4 = 4x^2 P_{ii}(1 - P_{ii})$$

$$f_5 = x(P_{ii} - x + xP_{ii}^2).$$

1.2 Oscillation strength and frequency of optical phonon

Behavior of optical phonon in an alloy semiconductor can be derived by a simple chain model which takes into account the effect of second neighbor atom. A model which includes the effect of atoms distributed in infinite length has been reported.[4] Frequency of the optical phonon from $Ga_{1-x}In_xAs$ can be explained quantitatively only in limited range of composition $0.2 \leq x \leq 1.0$. However the model can not explain one mode behavior of phonon in $Ga_{1-x}In_xAs$ ($0 \leq x \leq 0.2$). On the other hand, the linear chain model explain the frequency and mode strength of phonons from ternary alloy such as GaInAs, GaInSb, GaAsSb, InAsSb, AlInAs, AlInSb and AlInP over the entire range of composition as reported by Yamazaki et al.[2] Therefore we used the modified linear chain model. In this model three types of force constants, first nearest-neighbor force constants $k_{Ga-As}(i)$ between gallium and arsenic atoms and $k_{In-As}(i)$ between indium and arsenic atoms in unit i, and a second-nearest neighbor force constant k_{Ga-In} between gallium and indium atoms are taken into account. If we represent the atomic masses by m_{Ga}, m_{In} and m_{As} the equations of motion can be written as

$$m_{As}\ddot{u}_{As} = -(A + xk_4 - 4/3 \cdot \pi N x e'_{As} e'_{In})(u_{As} - u_{In})$$

$$- \{B + (1-x)k_5 - 4/3 \cdot \pi N(1-x)e'_{As}e'_{Ga}(u_{As} - u_{Ga}) - e_{As}E$$

$$(1-x)m_{Ga}\ddot{u}_{Ga} = -\{B + (1-x)k_5 - 4/3 \cdot \pi N(1-x)e'_{As}e'_{Ga}\}(u_{Ga} - u_{As})$$

$$- x(1-x)(k_3 + 4/3 \cdot \pi N e'_{Ga}e'_{In})(u_{Ga} - u_{In}) - (1-x)e_{Ga}E$$

(3)

$$xm_{In}\ddot{u}_{In} = -(A + xk_4 - 4/3 \cdot \pi N e'_{As}e'_{In})(u_{In} - u_{As})$$

$$- x(1-x)(k_3 + 4/3 \cdot \pi N e'_{Ga}e'_{In})(u_{In} - u_{Ga}) - xe_{In}E$$

$$A = \sum_{i=1}^{5} Z(i) f_i k_1(i)$$

$$B = \sum_{i=1}^{5} \{1 - Z(i)\} f_i k_2(i)$$

where $Z(i)$ is the fractional coefficient as reported later[2] and $\ddot{u}_i$, N_i, e'_i, e_i and E_i are accelerations of ions, numbers of ion pairs, local effective charges, effective charges and electric fields, respectively.[3] The resolution of these equations, that is temperature dependence of phonon frequencies and intensity ratio were obtained by solving the force-constant matrix element using computer M-200 (Fujitsu) as shown in Appendix. The temperature dependence of bulk modulus B' of binary semiconductors were used in the calculation. The relation between force constants and bulk modulus B' is represented by the following equation.[5]

$$k_{\text{Ga-As}} + P_{\text{gg}}k_{\text{In-As}} = 4aB' \tag{4}$$

k_i is a force constant and a is a lattice constant which varies with temperature as shown in Appendix. The value of the force constant for Ga-As mode in $\text{Ga}_{1-x}\text{In}_x\text{As}$ was determined such that temperature dependence of the force constant was proportional to that of GaAs single crystal. The relation between oscillator strength and phonon frequency for two modes in a ternary alloy is given by setting the dielectric function equal to zero assuming the damping coefficient is much smaller than 1[6] as shown in the following equations,

$$\omega_{L1} + \omega_{L2}^2 = (1 + I_1/\varepsilon_\infty)\omega_{T1}^2 + (1 + I_2/\varepsilon_\infty)_{T2}^2$$

$$I_2/I_1 = \{\varepsilon_\infty \omega_{L1}\omega_{L2}/\omega_{T1}\omega_{T2})^2 - 1)(1/I_1) - 1\} - 1 \tag{5}$$

where ω_{L1}, ω_{L2}, ω_{T1}, ω_{T2}, I_1, I_2, and ε_∞ are frequencies of longitudinal optical (LO) and transverse optical (TO) phonons, oscillator strengths of LO phonons for two vibrational modes, for example Ga–As and In–As modes, and dielectric constant at high frequency, respectively. Suffixs of 1 and 2 in ω_{Li}, ω_{Ti} and I_i show the two binaries constituting ternary alloys. When one kind of bond in the ternary alloy has tensile stress and the other has compressive one equation (5) is rewritten as follows.

$$I_2'/I_1' = \{\varepsilon_\infty((\omega_{L1} + \Delta\omega_{L1})^2(\omega_{L2} - \Delta\omega_{L2})^2/(\omega_{T1} + \Delta\omega_{T1})^2(\omega_{T2} - \Delta\omega_{T2})^2$$

$$- 1)(1/I_1) - 1\} - 1 \tag{6}$$

$\Delta\omega_{L1}$, $\Delta\omega_{L2}$, $\Delta\omega_{T1}$ and $\Delta\omega_{T2}$ are extra shifts from calculated results. I_1' and I_2' are the oscillator strength with stress. The oscillator strength of first order Raman spectra for GaSb, Ge and Si having stress in the range from 7×10^9 to 1×10^{10} dyn/cm^2 is almost identical to the strength of TO phonons for the stress free semiconductors within an deviation of 3% as reported by Cardona et al.[7] and Weinstein et al.[8] The fact stands also in the case of ternary alloys as will be discussed directly below for $\text{Ga}_{1-x}\text{In}_x\text{As}$ as an example. The TO phonon mode in $\text{Ga}_{1-x}\text{In}_x\text{As}$ ($x=0.53$) was obtained from (111)B samples, because TO phonons are sensitive to stress and can not be measured from a (100) surface according to the selection rule. A value of the intensity ratio I_2/I_1 was identical within about 0.25% to that calculated assuming the stress of 3×10^{10} dyn/cm^2 and data obtained from experiments listed in Table 1. That is, the relations $\Delta\omega_{L1}/\omega_{L1}\doteqdot\Delta\omega_{T1}/\omega_{T1}$, $\Delta\omega_{L2}/\omega_{L2}\doteqdot\Delta\omega_{T2}/\omega_{T2}$ are held in eq. (6). It means that change in β by 0.1 corresponds to that in stress of more than about 1×10^{10} dyn/cm^2. Shift of 1 cm^{-1} of phonon frequency which is within accuracy of measurement corresponds to change of the stress by 3×10^9 dyn/cm^2. Therefore, a clustering parameter can be estimated from the intensity ratio

Table 1. Phonon frequencies and oscillator strength of two phonons in $Ga_{0.47}In_{0.53}As$.

Mode		Lo (cm^{-1})	To (cm^{-1})	I
GaAs	Calculated	280	265	1.6
	Experiment	272	257.5	1.55[a]
InAs	Calculated	233	226	0.4
	Experiment	236	229	0.39[a]

[a]Calculated from $I'=(l-c)I$, where $c=0.03$.

between two modes while stress accumulated in a ternary is obtained from a frequency shift of LO phonon with a good accuracy.

2. Experimental Results and Discussion

2.1 Raman spectra of ternary alloys

Temperature dependence of Raman spectra from ternary alloys were measured using a quartz box which contained a heater as shown in Fig. 3. The quarts box was mounted in the sample room attached to a triple monochrometer model JRS-400T (JEOL). The resolution of the monochrometer was 0.5 cm^{-1} with a slit of 50 μm. Nitrogen gas was kept flowing in the quartz box

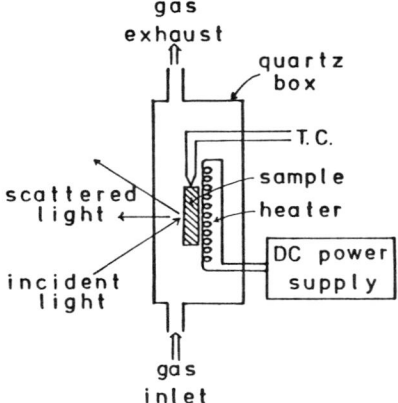

Fig. 3. Configuration of the sample holder used in the measurement of temperature dependence of Raman spectra.

during the measurement. Temperature of the sample was monitored by a thermocouple attached to the surface of the sample. A 514.5 nm line of Argon ion laser was used as an exciting source. Power density of the laser beam at the surface of the sample was kept low such as 6×10^2 W/cm^2 in order to give no effect on the local temperature of the surface. A temperature estimated from Stokes and anti-Stokes intensity ratio[9] was in ±20 degree of that measured by a thermocouple. Scattered light was collected in the backscattering geometry. Sensitivity of the photo-multiplier was calibrated by using a halogen lamp.

Four kind of ternary alloys $Ga_{1-x}Al_xAs$, $Ga_{1-x}In_xAs$, $Ga_{1-x}In_xP$ and $GaAs_{1-x}P_x$ were used as samples. Properties of the samples are listed in Table 2. Composition of the alloys were confirmed by X-ray diffraction method.

Raman spectra from $Ga_{0.47}In_{0.53}As$ in a range from 200 to 325 cm^{-1} at various temperatures are shown in Fig. 4. It is clear that the relative intensities between LO phonons based on GaAs and InAs modes changes with temperature. The peaks at about 272 and 273.5 cm^{-1} at about 291 K are the LO phonons based on Ga–As and In–As bonds, respectively. The peaks at 237.5 cm^{-1} is assigned to InAs mode LO phonon on many experimental data.[10,11] Theoretical analysis of lattice vibration for $Ga_{0.47}In_{0.53}As$ reported up to date[12,13] indicate also that $Ga_{0.47}In_{0.53}As$ has two LO phonons. Although Pearsall et al.[14] and Pinczuk et al.[15] reported that $Ga_{0.47}In_{0.53}As$ shows one mode behavior, their Raman spectra are likely to be misunderstood because intensity of InAs mode LO phonon is very weak if Raman spectra were measured at a relatively low temperature. Detailed discussion is given at Subsection 2.2.

Raman spectra from other ternary alloys listed in Table 2 essentially consist of two LO phonons in a range from 200 to 425 cm^{-1} at temperatures between 290 and 690 K. Another example for $Ga_{0.5}Al_{0.5}As$ is shown in Fig. 5.

Table 2. Properties of the ternary alloys used in the experiment.

Sample	$Ga_{1-x}Al_xAs$	$Ga_{1-x}In_xAs$	$Ga_{1-x}In_xP$	$GaAs_{1-x}P_x$
Composition x	0.50	0~0.53	0.48	0.30
Thickness (μm)	0.5	0.3~90	0.8	0.5
Substrate	GaAs	GaAs, InP	GaAs	GaAs
Carrier conc. (cm^{-3})	1.8×10^{16}	1.5×10^{16}	3×10^{16}	2×10^{18}
Growth method	LPE	LPE, VPE	LPE	LPE
Orientation	(100)	(100)	(100)	(100)

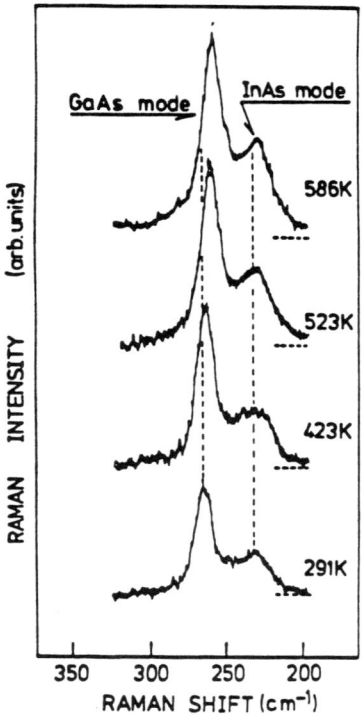

Fig. 4. Raman spectra at various temperatures from $Ga_{0.47}In_{0.53}As$.

In this paper, intensity ratios between two LO phonons and frequencies of the alloys are analyzed to estimate a clustering parameter and stress in each bond.

2.2 Estimation of a clustering parameter

Figure 6 shows temperature dependence of intensity ratios of LO phonons for $Ga_{0.47}In_{0.53}As$ and $Ga_{0.5}Al_{0.5}As$. Solid lines and dots are calculated and experimental results, respectively. The temperature dependence of the intensity ratio between two LO phonons was measured in order to make sure the calculated results including temperature dependence. The clustering parameter β of $Ga_{0.47}In_{0.53}As$ is estimated to be about 0.3 between 300 and 700 K from Fig. 6(a). The intensity ratio between GaAs and InAs modes at a low temperature is much larger than that at a high temperature. Therefore the data reported as one mode by Pearsall et al.[14] are likely to be

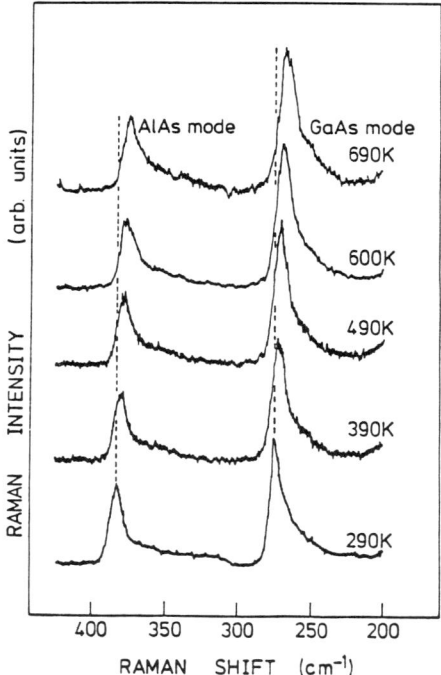

Fig. 5. Raman spectra at various temperatures from $Ga_{0.5}Al_{0.5}As$.

misjudged because they were obtained at 135 K by magneto-phonon resonance measurement.

Significant difference in a clustering parameter could not be observed between $Ga_{0.47}In_{0.53}As$ epitaxial layers grown by LPE and VPE methods.

The fact that experimental results are on the calculated curve for $\beta=0$ in the case of $Ga_{0.5}Al_{0.5}As$ as shown in Fig. 6(b) means that a clustering parameter is zero for the alloy. Clustering parameters obtained in this manner for ternary alloys listed in Table 2 are plotted in Fig. 7 as a function of relative difference in lattice constant ($\Delta a/\bar{a}$) between binaries which compose ternary alloys. In Fig. 7 calculated values of β for various ternary alloys which has a midrange of alloy composition, that is $x=0.5$, are also plotted. The calculated results were determined as a value at which the excess free energy of mixing takes the minimum in the regular solution model in which an effects due to clusters are taken into account.[17] Clustering parameters estimated by analyzing Raman spectra are very close to those calculated except for

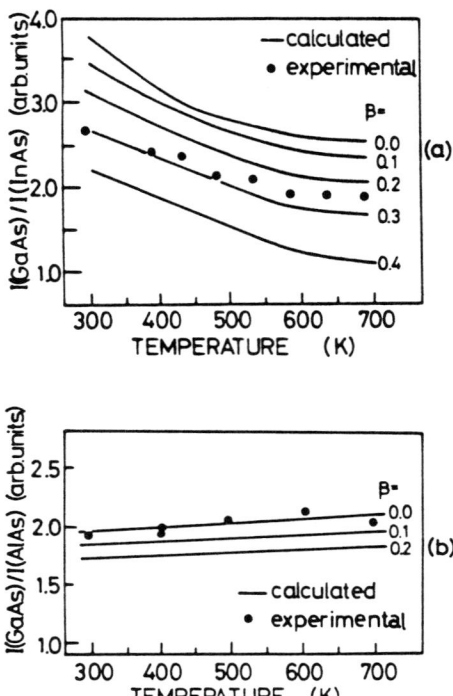

Fig. 6. Temperature dependence of intensity ratio between (a) GaAs and InAs modes in $Ga_{0.47}In_{0.53}As$, and (b) GaAs and AlAs modes in $Ga_{0.5}Al_{0.5}As$.

$GaAs_{0.7}P_{0.3}$ although the model of calculation is rather simple. The reason that of a clustering parameter $GaAs_{0.7}P_{0.3}$ obtained from Raman spectra is significantly larger than that calculated has not been made clear. One of the reason is considered that carrier concentration of $GaAs_{0.7}P_{0.3}$ is very high as listed in Table 2. When a polar semiconductor has a high carrier concentration its LO phonon intensity is weakened due to LO phonon-plasmon coupling.[16]

It is clear from Fig. 7 that a clustering parameter β in ternary alloy semiconductor increases with relative difference between lattice constants of two binary semiconductors ($\Delta a/\bar{a}$) which constitute the ternary alloy. The reason is considered, as Yamazaki et al.[17] have pointed out, that a ternary alloy which is formed by two kinds of binaries whose lattice constants are different significantly has a smaller excess free energy of mixing when some clusters are formed than when different kind of atoms are adjacent always.

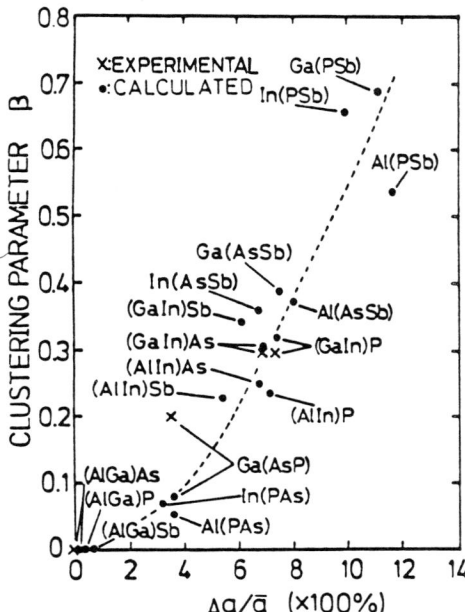

Fig. 7. Clustering parameters for various ternary alloys as a function of relative difference in lattice constants between constituing binaries.

2.3 Estimation of internal stress accumulated in each bond

At first, effects of lattice mismatch between an epitaxial layer and a substrate on stress are discussed. The lattice deformation based on lattice mismatch between a $Ga_{1-x}In_xAs$ ($0 \leq x \leq 0.5$) epitaxial layer and a GaAs substrate has been reported by Nagai.[18] The lattice deformation in the structure is directional, that is, the lattice of $Ga_{1-x}In_xAs$ is deformed tetragonally such that the lattice constant perpendicular to the epitaxial layer-substrate interface is larger than that parallel to the interface. According to the paper,[18] however, the lattice deformation is relaxed by the creation of misfit dislocation when the $Ga_{1-x}In_xAs$ epitaxial layer has a composition of $0.2 \leq x \leq 0.5$ and a thickness larger than 20 μm. The stress would be accumulated in the $Ga_{1-x}In_xAs$ epitaxial layer with a composition $0 \leq x \leq 0.2$. Then frequency shift of the optical phonon due to the stress based on lattice mismatch should change generally with the thickness of epitaxial layer because the stress decrease with thickness. However, the depth profile of the GaAs-LO phonon frequency from $Ga_{1-x}In_xAs$ ($0 \leq x \leq 0.11$) epitaxial layers grown on a GaAs substrates shows no change with the thickness of the epitaxial layer as shown in Fig. 8. The depth profile of composition obtained by electron probe micro analysis and that of the intensity of LO phonon are also shown. The sample of $Ga_{1-x}In_xAs$ ($0 \leq x \leq 0.15$) grown on a GaAs substrate composes of a compositionally graded layer of a thickness about 10 μm and a compositionally constant layer of about 80 μm. From these results, the extra shift of LO phonon frequency from the calculated one as shown in Fig. 7 is considered to be not due to the lattice mismatch but due to internal stress in $Ga_{1-x}In_xAs$. That is, the internal stress in $Ga_{1-x}In_xAs$ is not introduced by the substrate but the arrangement of atoms with different sizes. The direction of the stress can not be determined at present experimentally. We assume here that the stress is isotropic macroscopically. The extra shift of an optical phonon frequency due to stress is given by the following equation.[19]

$$\Delta\omega = (F/6\omega_0)(p + 2q)(S_{11} + 2S_{12}) \qquad (7)$$

$\Delta\omega$: extra frequency shift due to stress
F: stress
ω_0: frequency without stress
p, q: splitting parameters
S_{11}, S_{12}: elastic compliances

In order to estimate a value of the stress, temperature dependence of the extra shift was measured. Figure 9 shows temperature dependence of LO

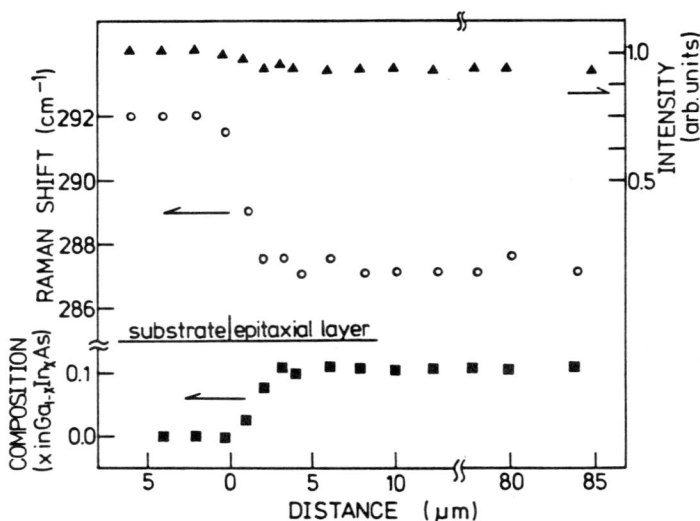

Fig. 8. Depth profile of phonon frequency, intensity and composition of In atom in $Ga_{1-x}In_xAs$ ($0 \leq x \leq 0.11$).

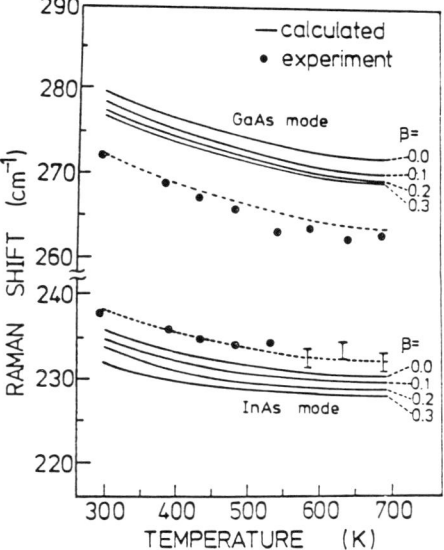

Fig. 9. Temperature dependence of phonon frequencies from $Ga_{0.47}In_{0.53}As$.

phonon frequencies based on GaAs and InAs modes in $Ga_{0.47}In_{0.53}As$. Solid lines and dots are calculated and experimental results, respectively. Temperature dependence curves of the phonon frequencies obtained experimentally are parallel to those calculated as shown in Fig. 9. It means that a phonon frequency changes with temperature according to that of force constants as shown in Appendix. The extra shift whose magnitude is independent to temperature is due to stress as reported by Evance et al.[20] Frequency of GaAs mode is lower than that calculated by eq. (6) with $\beta=0.3$ while frequency of InAs mode is higher than that calculated. The results indicate that stress accumulated in Ga–As bond is tensile and that in In–As bond is compressive. Extra shift of the frequency for GaAs mode is -5 cm^{-1} and that for InAs mode is $+6$ cm^{-1}. Internal stress calculated from the extra frequency shift using eq. (7) are 1.5×10^{10} dyn/cm^2 for Ga–As and In–As bonds, respectively. Internal stress estimated in this manner for other ternary alloys are plotted against $\Delta a/\bar{a}$ in Fig. 10. Physical constants used in the calculation are listed in Table 3. Internal stress accumulated in a bond of ternary alloy increases with that in $\Delta a/\bar{a}$. Stress accumulated in one bond corresponding to the shorter lattice constant of binary is due to tensile stress and that in the other bond corresponding to the longer lattice constant is due to compressive one.

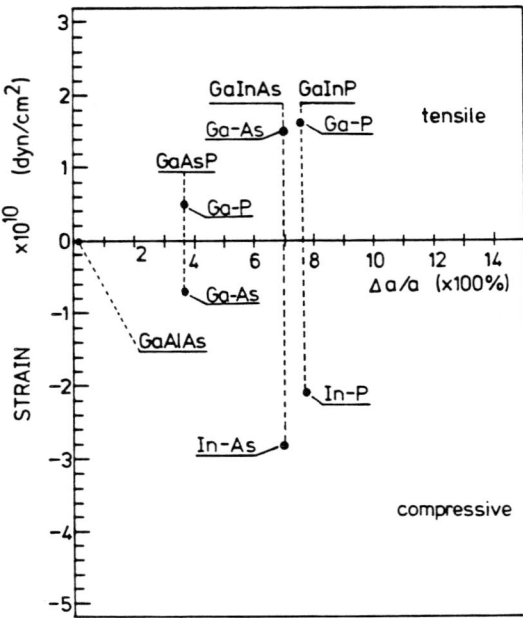

Fig. 10. Internal stress for each bond in ternary alloys as a function of relative difference in lattice constants between constituting binaries. $\bar{a}$: a middle values of lattice constants for two binaries constituting ternary alloys.

Table 3. Physical constants used in the calculation of internal strain energies and bond lengthes.

Binary	Elastic compliance $S_{11}+2S_{12}$ $(=10^{-12} \text{ cm}^2/\text{dyn})$	Dimensionless splitting parameter $(P+2q)/6w_0^2$	Bulk modulus B $(=10^{-12} \text{ cm}^2/\text{dyn})$
GaAs	0.445	0.90	1.326
InAs	0.575	0.85	1.727
GaP	0.376	0.9	1.127
InP	0.460	0.87	1.808
AlAs	0.435	0.9	1.115

2.4 Length of bond having stress

Accumulation of tensile type stress in a bond means that the bond length is larger than that without stress while accumulation of compressive type stress means that the bond length is shorter than that without stress. Length of a bond having stress can be calculated by the following equation assuming the stress to be isotorpic as mentioned in the previous section

$$l = (1/4)\sqrt{3}a(1+\varepsilon) \tag{8}$$

where a is a lattice constant and,

$$\varepsilon = (1/3)(S_{11} + 2S_{12})F \tag{9}$$

S_{11} and S_{12} are compliances and F is stress, respectively. The strain energy tensor $\vec{\varepsilon}$ with hydrostatic pressure is given by the following equation according to Martin's model.[21]

$$\vec{\varepsilon} = \varepsilon \begin{pmatrix} 100 \\ 010 \\ 001 \end{pmatrix} \tag{10}$$

Length of Ga–As bond and that of In–As bond in $Ga_{0.47}In_{0.53}As$ were estimated 2.47 and 2.59 Å, respectively. Bond lengths of Ga–As and In–As bonds in $Ga_{1-x}In_xAs$ were compared to those obtained with Extended X-ray Absorption Fine Structure (EXAFS) reported by J. Bellessa et al.[22] Figure 11 shows Raman shift of Ga–As mode in $Ga_{1-x}In_xAs$ with various values of x. Length of Ga–As bond corresponding to the Raman shift calculated from the stress is plotted in Fig. 12 against the composition x. Length of In–As bond in $Ga_{1-x}In_xAs$ can not be estimated with a good accuracy in a compositional range of $0 \leq x \leq 0.5$ because intensity of InAs LO phonon is very weak as known to be partly-two mode. Bond length-composition characteristics obtained with EXAFS are also plotted.[22] Values of bond length estimated from Raman shift and EXAFS are almost identical. It means that a value of stress estimated from extra Raman shift is reasonable. In addition, it is clear that bond length in $Ga_{0.47}In_{0.53}As$ does not obey Vegard's law microscopically as reported by Mikkelsen et al.[23] Bond lengths in other ternary alloys than $Ga_{1-x}In_xAs$ ($0 \leq x \leq 0.53$) listed in Table 4 could not be ascertained by comparison with other data because we could find no data such as EXAFS for those materials.

3. Summary

Estimation of a clustering parameter and stress accumulated in a bond of

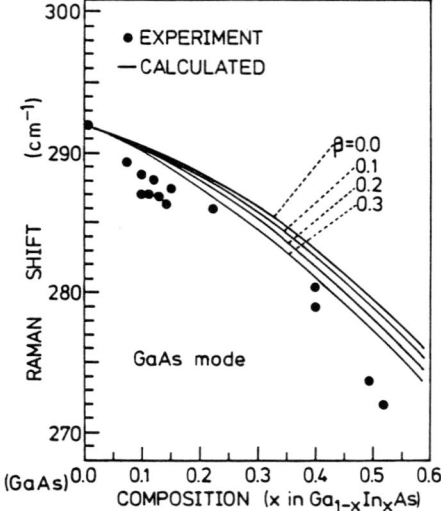

Fig. 11. Composition dependence vs frequency of GaAs mode optical phonon obtained from $Ga_{1-x}In_xAs$ ($0 \leq x \leq 0.53$).

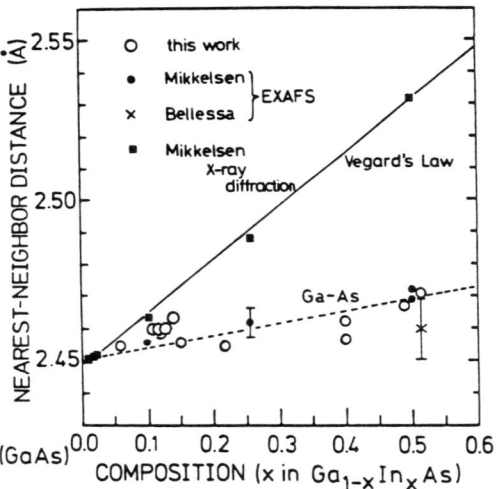

Fig. 12. Nearest-neighbor atomic distance in Ga–As bond for $Ga_{1-x}In_xAs$ ($0 \leq x \leq 0.53$) as a function of composition.

Table 4. Bond length derived from strain energies for each bond in ternary alloys. Values in the parentheses are bond lengths in binaries.[26]

Alloy	x	Bond length for ternary (Å)		Vegard's low
		Shorter bond	Longer bond	
$Ga_{1-x}Al_xAs$	0.50	[Ga-As] 2.45 (2.45)	[Al-As] 2.45 (2.45)	2.45
$GaAs_{1-x}P_x$	0.30	[Ga-P] 2.37 (2.36)	[Ga-As] 2.44 (2.45)	2.41
$Ga_{1-x}In_xAs$	0.53	[Ga-As] 2.47 (2.45)	[In-As] 2.59 (2.62)	2.54
$Ga_{1-x}In_xP$	0.48	[Ga-P] 2.37 (2.36)	[In-P] 2.51 (2.54)	2.44

a ternary alloy from Raman spectra was proposed. Clustering parameters estimated from analysis of Raman spectra for $Ga_{0.5}Al_{0.5}As$, $Ga_{0.52}In_{0.48}P$ and $Ga_{0.47}In_{0.53}As$ were almost identical to those, respectively, obtained by theoretical calculation of excess free energy of mixing. Although no data measured by other methods were found to be compared with stress estimated from Raman spectra directly, bond lengths of Ga–As and In–As bonds in $Ga_{1-x}In_xAs$ corresponding to the stress were almost identical to those estimated by EXAFS.

APPENDIX

The relation between bulk modulus and elastic compliances is written as

$$B' = (1/3)(C_{11} + 2C_{12}) \tag{A1}$$

The compliances C_{11} and C_{12} have temperature dependences. The data of C_{11}, C_{12} and C_{44} for binary III–V semiconductors have been reported[24,25] and used in the calculation described in this paper. We can obtain the temperature dependence of force constant from eq. (A1) using eq. (4).

The temperature dependence of a lattice constant can be obtained from the following equation,

$$a(T) = a_0(1 + \alpha T) \tag{A2}$$

where a_0 is a lattice constant at room temperature and α is linear expansion coefficient.

II. CHARACTERIZATION OF DISORDERED BONDS IN Si-IMPLANTED $Ga_{0.47}In_{0.53}As$

1. Experiment

The samples used in the experiment were $Ga_{0.47}In_{0.53}As$ epitaxial layers grown on Fe-doped (100) and (111)B InP substrates by liquid phase epitaxy. Carrier concentration of the epitaxial layers was about 2×10^{16} cm^{-3} at room temperature before ion implantation. The thickness of the layers was about 4 μm. The $Ga_{0.47}In_{0.53}As$ epitaxial layers were implanted with Si at 300 keV with a dose of 1×10^{14} cm^{-2} at room temperature. The (100) and (111)B substrates were positioned so that the <100> and <111>B axes are about 10° off from the ion beam direction, respectively. The implanted samples were annealed with no encapsulant at temperatures from 500 to 700°C for 20 min in an atmosphere of flowing nitrogen gas. Raman spectra were measured in a backscattering geometry at room temperature with a 514.5 nm line of Ar$^+$ laser. Polarization characteristics of Raman spectra were measured by using two polarizers against incident and scattered lights. The Γ_{15} component of lattice vibration was measured in a relation so that incident and scattered lights were polarized parallel to the {100} direction and perpendicular to the incident plane, respectively. The Γ_1 component was obtained in a relation such that both incident and scattered light were polarized parallel to the {100} direction.

2. Results and Discussion

Raman spectra from unimplanted and implanted $Ga_{0.47}In_{0.53}As$ are shown in Fig. 13. Spectra denoted by A and B obtained from $Ga_{0.47}In_{0.53}As$ grown on (111)B InP substrates. Raman spectra denoted by A, C and B, D were obtained from unimplanted and implanted samples, respectively. Two sharp peaks corresponding to LO and TO phonons based on GaAs and InAs modes, respectively, and peaks lower than 200 cm^{-1} based on acoustic phonon[10] are observed before implantation for (100) surface. The GaAs TO mode is observed also for (111)B surface due to selection rule.[27] A new peak corresponding to InAs TO mode can't be seen, however, for the (111)B surface because LO and TO InAs modes in $Ga_{0.47}In_{0.53}As$ have almost the same wave numbers.[2] Assignment of the optical phonon has been reported.[2]

A sharp peak at about 240 cm^{-1} based on InAs optical modes obtained from the unimplanted samples is not observed in the spectra from the implanted samples. On the other hand, a peak at about 270 cm^{-1} based on GaAs optical phonon can be observed before and after implantation. Increase in intensity of peaks from 200 to 240 cm^{-1} is observed on the spectra from

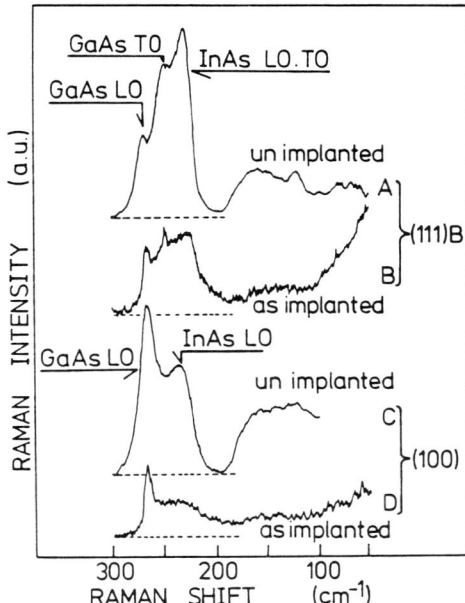

Fig. 13. Raman spectra in a spectral range of 50–300 cm^{-1} from unimplanted and implanted Ga$_{0.47}$In$_{0.53}$As.

implanted samples. It has been reported that intensity of the optical phonon decreased and that new peaks in a region from 220 to 270 cm^{-1} appeared in the spectra from Si-implanted GaAs. Therefore, it is possible that the peaks in 200–240 cm^{-1} described above consist of those based on an amorphous phase of GaAs and InAs. However, the difference in presence of a sharp peak due to the optical phonons between GaAs and InAs for implanted Ga$_{0.47}$In$_{0.53}$As indicates that disorder is introduced more easily into InAs sublattice than into GaAs sublattice in Ga$_{0.47}$In$_{0.53}$As independent of irradiation angles. The difference can be considered due to the difference in cohesive energies between GaAs and InAs.[28]

In order to make clear the difference in degree of deviation from zincblende structure introduced by implantation between Ga–As and In–As bonds, polarization characteristics of the Raman spectra were measured. The relation between annealing temperature and relative intensity of Γ_{15} to Γ_{1}, $I(\Gamma_{15})/I(\Gamma_{1})$ for GaAs and InAs modes in Ga$_{0.47}$In$_{0.53}$As,[10] respectively, are shown in Fig. 14. Γ_{15} and Γ_{1} components correspond to the symmetries of lattice vibration allowed and forbidden to the zincblende structure, respectively. Therefore, a small value of the ratio means a presence of deviation from

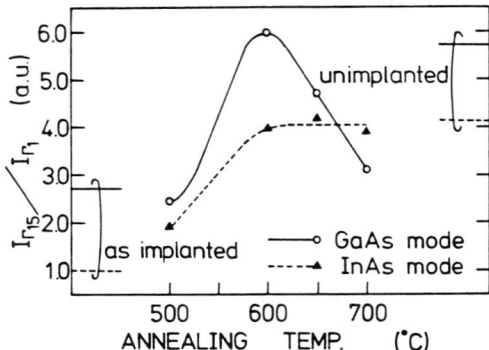

Fig. 14. Relation between annealing temperature and intensity ratio between Γ_{15} and Γ_1 components from Si-implanted $Ga_{0.47}In_{0.53}As$.

the zincblende structure. That is, the presence of deviation from the zincblende structure in Ga–As bond is clear from Fig. 14 although it cannot be said from the spectra shown in Fig. 13.

Each ratio for both GaAs and InAs modes increases with annealing temperature and recovers at 600°C to that before implantation, respectively, as shown in Fig. 14. Spectra obtained at various stages of annealing for (100) surface are shown in Fig. 15. A peak of InAs LO mode appears again with increase in annealing temperature. This reappearance is coincident with increase in the ratio $I(\Gamma_{15})/I(\Gamma_1)$ for InAs mode. The results mean that the disorder in both Ga–As and In–As bonds is removed simultaneously even though the degree of disorder is different between them.

The ratio $I(\Gamma_{15})/I(\Gamma_1)$ for GaAs mode decreases again with annealing at a temperature higher than about 650°C as shown in Fig. 14. It is due to decrease in LO phonon intensity, as shown in Fig. 15, because the LO phonon has Γ_{15} symmetry and couples with a plasmon when the semiconductor has a high carrier concentration. In general, LO phonon intensity decreases when it couples with a plasmon while another plamon mode which has almost the same wave number as the TO phonon[16] increases with increase in carrier concentration. Therefore, the intensity ratio for InAs mode does not decrease by annealing at a temperature higher than 650°C because one of the plama modes overlaps with the LO phonon in the case of InAs as previously described. Carrier concentration of the implanted layer after annealing at 700°C was about 1×10^{18} cm^{-3}.

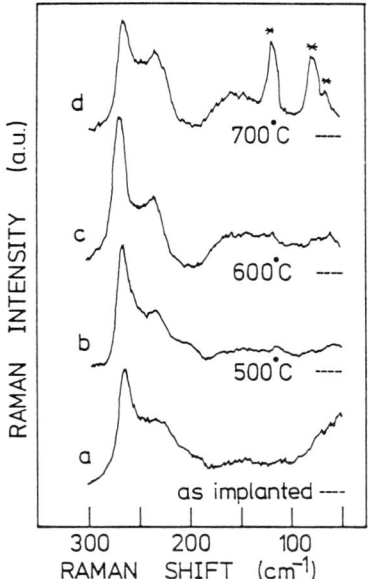

Fig. 15. Raman spectra of Si-implanted $Ga_{0.47}In_{0.53}As$ at various annealing temperatures. (*) laser lines.

3. Summary

It was found with laser Raman spectroscopy that disorder is introduced more easily into In–As bonds than into Ga–As bonds in $Ga_{0.47}In_{0.53}As$ by ion implantation. However, the disorder is removed simultaneously from both In–As and Ga–As bonds with annealing at a temperature higher than 600°C.

REFERENCES

1) N. A. Bert, A. T. Gorelenok, A. G. Dziganov, S. G. Konniku, V. N. Mdivani, I. S. Tarasov, and S. Ushikov: Sov. Phys. Semicond. **16** (1982) 35.
2) S. Yamazaki, A. Ushirokawa, and T. Katoda: J. Appl. Phys. **51** (1980) 3722.
3) H. W. Verleuer and A. B. Berker, Jr.: Phys. Rev. **149** (1966) 715.
4) G. M. Zinger, M. A. Il'in, E. P. Rashevskaya, and A. I. Ryskin: Sov. Phys. Solid State **21** (1979) 1522.
5) M. Born and K. Huang: *Dynamical Theory of Crystal Lattices*, ed. M. Cardona (Clarendon Press, Oxford, 1964) p. 111.
6) A. S. Berker: Phys. Rev. **136** (1964) A1290.
7) B. A. Weinstein and M. Cardona: Phys. Rev. **B5** (1972) 3120.
8) B. A. Weinstein and G. J. Piermarini: Phys. Rev. **B12** (1975) 1172.

9) R. Hart, R. L. Aggarwal, and Benjamin Lax: Phys. Rev. **B15** (1970) 638.
10) K. Kakimoto and T. Katoda: Appl. Phys. Lett. **40** (1982) 826.
11) K. Kakimoto and T. Katoda: Appl. Phys. Lett. **42** (1983) 811.
12) C. Pickering: J. Electron. Mater. **10** (1981) 901.
13) P. Kleinert: Phys. Status Solidi **b114** (1982) 459.
14) T. P. Pearsall, R. Carles, and J. C. Portal: Appl. Phys. Lett. **42** (1983) 436.
15) A. Pinczuk, J. M. Worlock, R. E. Nahory, and M. A. Pollak: Appl. Phys. Lett. **33** (1978) 461.
16) M. V. Klein: *Light Scattering in Solids*, ed. M. Cardona (Springer-Verlag, Berlin, Heidelberg, New York, 1975) Vol. 8, p. 159.
17) S. Yamazaki, M. Kishi, and T. Katoda: Phys. Status Solidi **b113** (1982) 421.
18) H. Nagai: J. Appl. Phys. **45** (1974) 3789.
19) F. Cerdeira, C. J. Buchenauer, F. H. Pollak, and M. Cardona: Phys. Rev. **B5** (1972) 580.
20) D. J. Evance and S. Ushioda: Phys. Rev. **B9** (1974) 1638.
21) R. M. Martin: Phys. Rev. **B1** (1970) 4005.
22) J. Bellessa, C. Gors, P. Launois, M. Quillec, and H. Launois: *Tenth International Symposium on Gallium Arsenide and Related Compounds, 1982*, ed. G. E. Stillman (The Institute of Physics, Bristol and London, 1983) p. 529.
23) J. C. Mikkelsen, Jr. and J. B. Boyce: Phys. Rev. Lett. **49** (1982) 1412.
24) C. W. Garland and K. C. Park: J. Appl. Phys. **33** (1962) 759.
25) R. Reifenberger, M. J. Meck, and J. Trivisonno: J. Appl. Phys. **40** (1969) 5403.
26) U. Piesbergern: *Heat Capacity and Debye Temperatures, Semiconductors and Semimetals*, ed. R. K. Willardson (Academic Press, New York, 1966) Vol. 2, p. 49.
27) R. Loudon: Adv. Phys. **13** (1964) 423.
28) J. C. Phillips: *Bonds and Bands in Semiconductors* (Academic Press, New York, 1973) p. 49.

ELECTRICAL PROPERTIES OF DX CENTER IN SELECTIVELY DOPED AlGaAs/GaAs HETEROSTRUCRURE

Masahiko TAKIKAWA and Masashi OZEKI

Fujitsu Ltd., 10-1 Morinosato-Wakamiya, Atsugi 243-01, Japan

Abstract. Energy structure of DX center was determined from a selectively Si doped $Al_{0.3}Ga_{0.7}As/GaAs$ heterostructure grown by molecular beam epitaxy. By using deep-level-transient-spectroscopy (DLTS) technique, drain current transient after gate pulse was measured for a long gate high-electron-mobility-transistor biased in a linear region. This technique enabled us to record DLTS-like spectra not only for the electron emission process but also for the electron capture process of the DX center. From the analysis of these spectra, we found that capture and emission activation energies had distributions with wide and narrow band-widths, respectively.

1. Introduction

It is well known that highly doped n-type AlGaAs presents a persistent photo-conductivity (PPC) effect which is attributed to electrons emitted from DX centers introduced by donor impurity doping.[1,2] Concentration and the other thermal properties of the DX center in a highly doped AlGaAs layer has not been well known, since a capacitance deep-level-transient-spectroscopy (DLTS) technique[3] is inadequate to detect the DX center using highly doped AlGaAs layer. DLTS peak due to DX center is distorted in the highly doped AlGaAs layer, because the junction capacitance is determined by DX center itself. Therefore, we cannot make quantitative characterization of DX center by using the capacitance DLTS spectrum.

Recently Valois *et al.* have detected the DX center from the selectively doped heterostructure.[4] By using the DLTS technique, the drain current transient after a positive gate bias pulse was measured for a long gate HEMT biased in a saturation region. We applied the similar technique to a long gate HEMT biased in a linear region. We have for the first time recorded the DLTS-like spectrum for the electron capture process by replacing the positive gate bias pulse with a negative pulse. From the detailed analysis of DLTS-like spectra recorded both for the electron emission and capture processes, we found that the capture and emission activation energies had distributions with wide and narrow band-widths, respectively. The origin of these broadening situations is discussed.

2. Experimental Procedure[6]

To record the spectrum for the electron emission process, applied gate voltage is changed as shown in the right hand side of Fig. 1(a). The corresponding band diagram of the selectively doped heterostructure is shown in the left hand side of the figure. Then the positive bias pulse is applied as shown in the middle of the figure. In the bias pulse period, empty deep levels located below imref are filled by electrons. As shown in the bottom part of the figure, deep level returns to be empty of electrons after the bias pulse is removed. Accordingly, the amount of interface charge increases. Then the interface current increases. The interface current difference between the gating time t_1 and t_2 is measured as a function of temperature.

As shown by Fig. 1(b), we can record the spectrum for the electron capture process, by replacing the positive pulse with a negative pulse. In the bias pulse period, filled deep levels located above imref emit electrons. After the bias pulse is removed, deep levels again become full of electrons, accordingly, the interface charge decreases. Therefore, the DLTS-like spectrum can be recorded by applying the DLTS technique to this decreasing interface current transient.

3. Sample Preparation

The selectively doped AlGaAs/GaAs heterostructure consisting of consecutively an undoped GaAs, a 60 Å undoped $Al_{0.3}Ga_{0.7}As$ layer and a 600 Å n-type $Al_{0.3}Ga_{0.7}As$ layer doped with Si to a level of 1.0×10^{18} cm^{-3} was grown on (100) GaAs semi-insulating substrate by MBE. The long gate HEMT was fabricated on this heterostructure. An active layer isolation was done by mesa definition and etching about 2000 Å. Then the Au-Ge/Au ohmic contact was made by an evaporation followed by an annealing for 3 min at 450°C in an N_2 atmosphere. A 300 Å thick Al film was deposited onto AlGaAs as a gate electrode. The gate width and length were 50 and 400 μm, respectively, and the source-gate and gate-drain spacings were 5 μm.

4. Results and Discussions

Figure 2(a) shows the temperature dependence of the DLTS-like spectra recorded for the electron emission process, where t_2/t_1 is kept at 10 with t_1 cranging from 0.02 to 2 ms. A stationary gate bias voltage, V_G, is −0.2 V.

Electrical Properties of DX Center 135

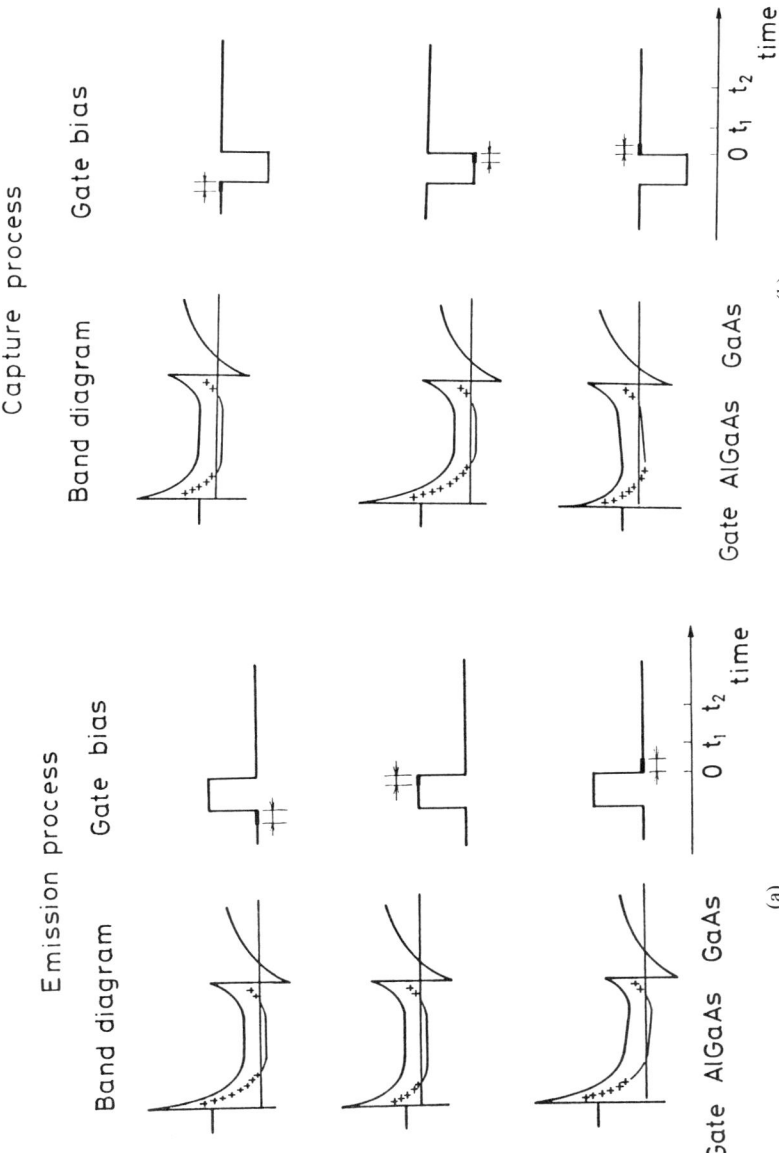

Fig. 1. Measurement procedure for the electron emission process (a) and that for the electron capture process (b).[6] The left hand side of the figure shows the applied gate bias voltage and the right hand side of the figure shows the corresponding band diagram under the gate region. Upper: Before bias pulse. Middle: Bias pulse duration. Lower: After bias pulse.

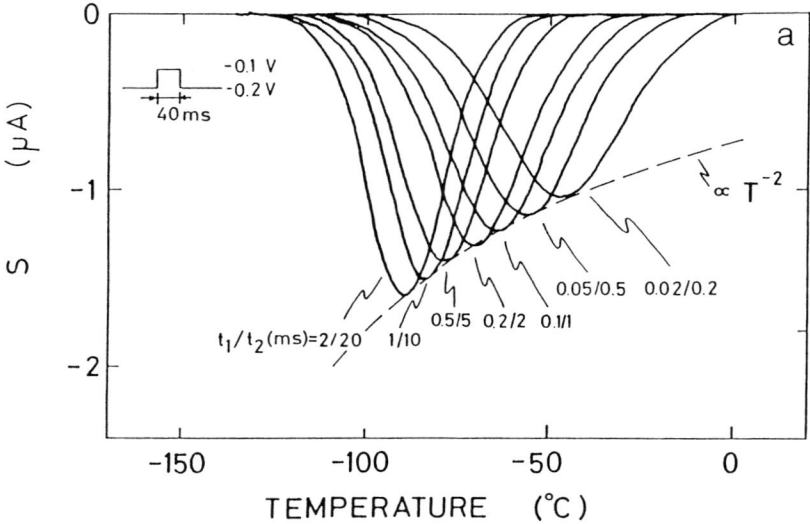

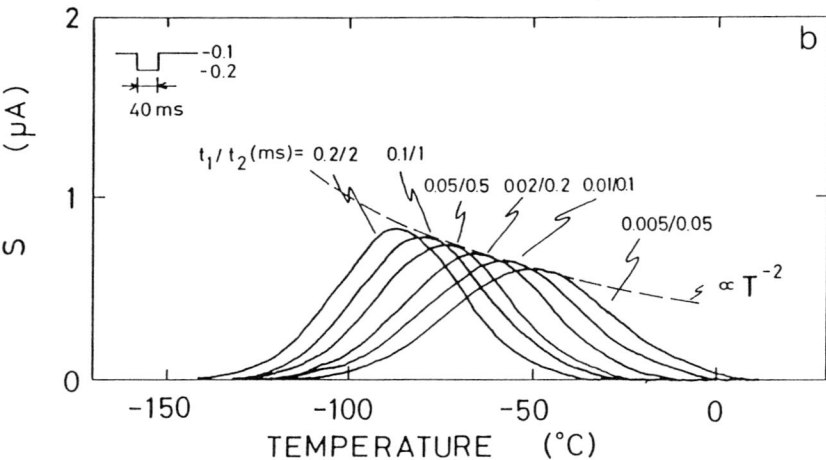

Fig. 2. Temperature dependence of the DLTS-like spectra (thick solid lines).[6] t_2/t_1 is held at 10. Dashed lines show that the peak height of the spectrum decreases in proportional to T^{-2}. (a) Spectra recorded for the electron emission process. t_1 ranging from 0.02 to 2 ms. (b) Spectra recorded for the electron capture process. Rate window ranging from 0.005 to 0.2 ms.

Then, the pulse bias with an amplitude of 0.1 V and width of 40 ms is repetitively applied to the gate. V_D of 0.1 V, which is in the linear region of output characteristics of our HEMT at temperatures from 77 K to room temperature, is applied. The peak heights of the spectra decreases with decreasing t_1, even if the ratio t_2/t_1 is kept constant. Since μ decreases in proportional to T^{-2} in this heterostructure above 100 K, μ is dominantly determined by the polar optical phonon scattering in this temperature range. Indeed the peak heights of the spectra decrease in proportional to T^{-2} as shown by dashed line in Fig. 2(a).

Figure 2(b) shows the temperature dependence of the DLTS-like spectra recorded for the electron capture process, where t_2/t_1 is kept at 10 with t_1 ranging from 0.005 to 0.2 ms. V_G is -0.1 V. The pulse bias with amplitude of -0.1 V and duration of 40 ms is repetitively applied to the gate. V_D is 0.1 V. The peak height of the spectra also decrease in proportional to T^{-2} as shown by dashed line in Fig. 2(b).

The approximated values of parameters were then determined by using the same method as is used in the conventional DLTS. Arrhenius plots shown by open circles in Fig. 3 were obtained from the corrected spectra for the electron emission process by using the same method as is used in the conventional DLTS. These plots fall on a straight line. The slope of this line

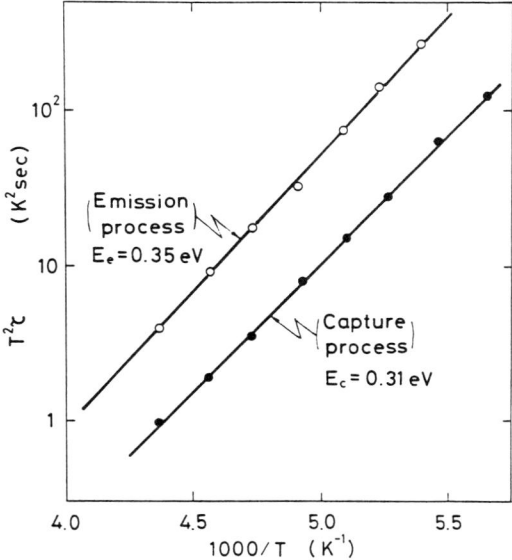

Fig. 3. Arrhenius plots obtained from the spectra recorded for the electron emission process (open circles) and for the electron capture process (filled circles).[6]

yields an emission activation energy, E_n, of 0.35 eV and a capture cross section at infinite temperature, σ_∞, of 5.0×10^{-14} cm^2. Arrhenius plots shown by the filled circles in Fig. 3 were obtained from the corrected spectra for the electron capture process. These plots also fall on a straight line. The slope of this line yields a capture activation energy, E_c, of 0.31 eV.

An extraordinarily wide half width of about 40 K was observed in the electron capture spectra. If several different levels with close activation energy are located, a wide half width of DLTS spectrum is expected. However, apparent single peak obtained for the large t_2/t_1 would sometimes separate into several peaks for the small t_2/t_1. Even in the spectra recorded for $t_2/t_1 = 2$, we never saw such double peaks as Lang et al. have reported for Te-doped AlGaAs.[2] On the contrary, an ordinary half width of about 25 K was observed in the electron emission spectra. There also appears single peak in the spectrum recorded for $t_2/t_1 = 2$. From these results, we conclude that the peak in the spectrum arises from only one level whose capture activation energy is distributed, while whose emission activation energy is not. This means that both E_T and W are distributed and electrons emitted from deeper level jump over lower barrier height.

The DLTS-like spectra for the electron emission and capture processes were also measured for the DX centers located at other positions and depths on this wafer. Half widths of any set of spectra were the same as those obtained from Fig. 2. Therefore, we conclude that the observed distribution arises from random fluctuations in the surrounding environment of the DX center cell. Summary of the electronic transition properties of the DX center in Si-doped $Al_{0.3}Ga_{0.7}As$ layer is shown in Fig. 4.

We discuss the origin of the broadening under random fluctuation. The electronic transition properties shown in Fig. 4 is explained in terms of a simple model of the configuration coordinate diagram depicted in Fig. 5, if we assume that the electronic state much fluctuates, while the spring constant and

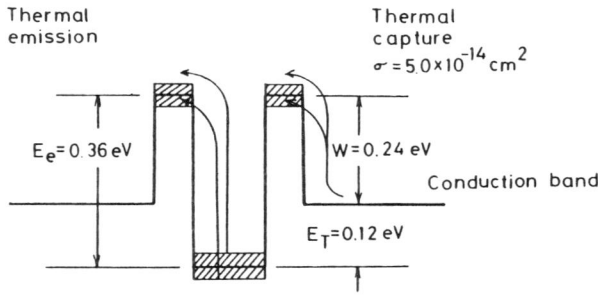

Fig. 4. Summary of the experimentally determined energy structure of the DX center in Si doped $Al_{0.3}Ga_{0.7}As$ layer grown by MBE.

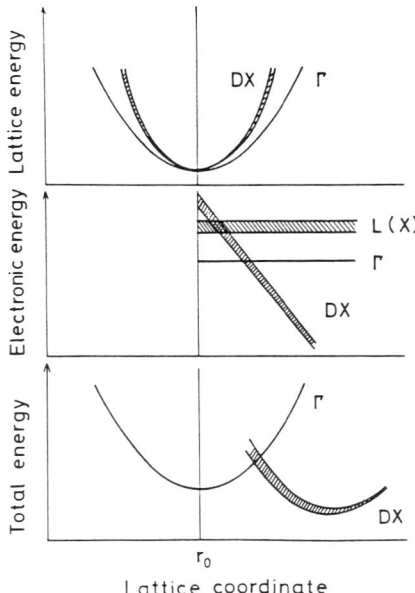

Fig. 5. Configuration coordinate diagram constructed to qualitatively fit the data for the Si related DX center in the $Al_{0.3}Ga_{0.7}As$ layer grown by MBE.[6] The value of spring constant of the Si-related DX center cell is assumed to be larger than that of the perfect crystal.[5]

the splitting parameter do not in the defect cell. It was reported that the approximate binding energy of the DX center follows the band edge of the higher lying band of large density of state primary associated with the X or L valley, which increases with Al mole fraction in the direct transition region.[2] This indicates that the energy difference between conduction band minimum and the electronic energy level of the DX center changes with the Al mole fraction in the surrounding atoms including not only the nearest neighbor but also other neighbors. More detailed results and discussions were described in ref. 6

5. Conclusion

By using DLTS technique, the drain current transient after the gate bias pulse was measured for a long gate HEMT biased in a linear region. This method enabled us to detect the DX center directly using the selectively doped heterostructure. From the analysis of the DLTS-like spectra recorded both for the electron emission process and the electron capture process of DX center, we found that the capture activation energy and the emission activation energy had distributions with wide and narrow band widths, respectively. These results can be understood by large fluctuation of the electronic state and small fluctuation of the spring constant and the splitting parameter in the DX center cell.

Acknowledgment

Part of this study was performed through Special Coordination Funds for Promoting Science and Technology of the Japanese Government.

REFERENCES

1) D. V. Lang and R. A. Logan: Phys. Rev. Lett. **39** (1977) 635.
2) D. V. Lang, R. A. Logan, and M. Jaros: Phys. Rev. **B19** (1979) 1015.
3) D. V. Lang: J. Appl. Phys. **45** (1974) 3023.
4) A. J. Valois and G. Y. Robinson: IEEE Trans. Electron Device Lett. **EDL-4** (1983) 360.
5) D. V. Lang and R. A. Logan: *Physics of Semiconductors, 1978* (Inst. Phys., Bristol and London, 1979) Conf. Ser. 43, p. 433.
6) M. Takikawa and M. Ozeki: Jpn. J. Appl. Phys. **24** (1985) 303.

PART 2: SILICON

POINT DEFECTS AND IMPURITIES IN SILICON CRYSTALS

Jun-ichi CHIKAWA

Photon Factory, National Laboratory for High Energy Physics, Oho-mach, Tsukuba-gun, Ibaraki 305, Japan

1. Introduction

To seek the future directions in materials and processing technology for VLSI, many investigations have been concentrated on intrinsic point defects, behavior of impurity oxygen, microdefects, and impurity effects on their formation in dislocation-free Si crystals.[1] Even though equilibrium concentrations of intrinsic defects at high temperatures are very low compared with the impurity concentration level in Czochralski-grown (CZ) crystals (see Table 1),[2-6] they play an important role in impurity complexing and microdefect formation which will be critical problems to be overcome when achieving a high degree of integration over a large area.

Studies on point defects are classified into three types: radiation-induced defects,[7] measurement of self and impurity diffusion,[8] and observation of agglomerates of point defects during crystal growth.[9] Defect behavior at low (radiation damage) and high temperatures has been investigated independently each other; defect behavior near equilibrium concentrations at high temperatures is essentially different from that of defects introduced as radiation damage at very high concentrations. Since intrinsic defects have a strong interaction with trace impurities remaining in Si crystals, intrinsic nature of point defects should be investigated with very pure crystals. For example, microdefects named "D-defects"[10] can be observed only with low carbon concentrations in undetectable levels (≤ 0.5 ppm) which were recently achieved.[11]

In this paper, recent investigations on microdefects introduced during crystal growth will be reviewed; current theories on coexistence of self-interstitials and vacancies will be described briefly. Recently, very small defects were found to distribute in float-zone (FZ) crystals free from swirl defects (interstitial-type) and D-defects (vacancy agglomerates).[11] By showing formation of this type of defects that depends upon formation of

Table 1. Equilibrium concentrations, $C_{\text{IO}}^{\text{eq}}$, $C_{\text{VO}}^{\text{eq}}$ (cm^{-3}), and diffusivities, D_{IO}, D_{VO} (cm^2/s), of self-interstitials and vacancies in silicon.

Ref.	Voronkov (2)	Seeger (3)	Inoue (4)	Yoshida (5)	Masters (6)
$C_{\text{IO}}^{\text{eq}}$	1.3×10^{14}	2×10^{16}	6×10^{17}		
D_{IO}	5×10^{-4}	4.3×10^{-6}	$\sim 10^{-7}$		
$C_{\text{VO}}^{\text{eq}}$	1.4×10^{14}			$\sim 10^{16}$	6.5×10^{14}
D_{VO}	3.5×10^{-4}			$\sim 10^{-5}$	4.2×10^{-4}

D-defects, a principal role of vacancies at high temperatures will be emphasized.

2. Interstitial-Vacancy Coexistence Models

Microdefects were often observed in swirl patterns in round slices of float-zoned (FZ) and Czochralski-grown (CZ) crystals. In longitudinal slices, they appeared in striated distribution. These defects have been referred to as "swirls" or "swirl defects". Swirl defects are classified into two kinds, A and B defects. A-defects were identified to be interstitial-type dislocation loops by electron microscopy in 1975.[12] Since then, the defects have been considered to be formed by condensation of self-interstitials supersaturated in cooling processes during growth, and origins of self-interstitials have been questioned. Petroff and de Kock[13] tried to interpret by "non-equilibrium interstitial model" in which self-interstitials are incorporated into the crystal in non-equilibrium concentrations. While, Seeger et al.[3] proposed "equilibrium interstitial model" in which the self-interstitials are assumed to be present as "extended interstitials", dominating over vacancies in equilibrium near the melting point. In the basis of *in-situ* X-ray observation on melting and growth processes, the present author[14] proposed a model of drop formation near the interfaces which leads to a local enhancement of interstitial concentrations. Recently, a coexistence model of vacancies and interstitials was proposed to explain oxidation-enhanced and retarded diffusion.[15] After Roksnoer and Van den Boom[10] found D-defects (vacancy agglomerates) in crystals grown at higher growth rates, many investigators have accepted the equilibrium coexistence of vacancies V and interstitials I at high temperatures.

From this viewpoint, Voronkov[2] attempted to explain formation of swirl and D-defects. He calculated concentrations C_V and C_I of vacancies and interstitials during the cooling process of crystals having both the defects at the equilibrium concentrations C_{VO}^{eq} and C_{VO}^{eq} at the melting point T_O. His theory starts from the two basic equations: From the condition that the recombination and pair-formation rates balance in the crystal (time required for recombination is much less than the cooling time), we have

$$C_I C_V = C_I^{eq} C_V^{eq}. \tag{1}$$

The flux of excess atoms into the crystal is equal to the difference of the interstitial and vacancy flux;

$$J = (vC_I - D_I \frac{\partial C_I}{\partial z}) - (vC_V - D_V \frac{\partial C_V}{\partial z}) \tag{2}$$

where z is the distance from the interface, v is the growth rate, and D_I and D_V are the diffusivities of vacancies and interstitials which are assumed to be nearly constant at high temperatures. By solving these equations, $C_I - C_V$ in the cooling crystal is obtained as a function of $v/G_O = \xi$ where G_O is the temperature gradient near the interface. Depending upon whether $C_I - C_V < 0$ or $C_I - C_V > 0$, D-defects or swirl defects are generated. According his results obtained using various experimental data, vacancies have a little higher concentration than interstitials at the melting point [$C_{IO}^{eq} < C_{VO}^{eq}$, $C_{VO}^{eq}/C_{IO}^{eq} = 1.07$ at T_O], but the diffusivity of interstitials is larger than that of vacancies ($D_V/D_I = 0.71$, $D_I C_{IO} > D_V C_{VO}$); consequently, for small values of ξ, interstitials diffuse dominantly into the crystals so that we have $C_I > C_V$, whereas when ξ is large both defects are incorporated with concentrations nearly equal to the equilibrium ones at T_O, and therefore $C_I < C_V$ ($C_I - C_V \rightarrow C_{IO}^{eq} - C_{VO}^{eq} < 0$ by increasing ξ).

In his theory, the equilibrium concentrations of C_I^{eq} and C_V^{eq} in the cooling process ($T < T_O$) were put to be zero. This approximation results in enhancing of up-hill diffusion toward the interface.

Tan et al.[9] introduced two equations for the steady state growth:

$$\frac{d}{dz}[D_I \frac{\partial (C_I - C_I^{eq})}{\partial z}] - v\frac{dC_I}{dz} - k_r(C_I C_V - C_I^{eq} C_V^{eq}) = 0. \tag{3}$$

$$\frac{d}{dz}[D_V \frac{\partial (C_V - C_V^{eq})}{\partial z}] - v\frac{dC_V}{dz} - k_r(C_I C_V - C_I^{eq} C_V^{eq}) = 0. \tag{4}$$

Since $C - C^{eq} = 0$ at the melting point T_O, when $C - C^{eq} > 0$ inside the crystal ($T < T_O$), the first term expresses the out-diffusion of point defects from the interface. The third term gives the difference the recombination and pair-generation rates (k_r: proportional constant). Although these equations have not been solved, Tan et al.[9] envisaged that $C_I^{eq} > C_V^{eq}$ in the temperature range

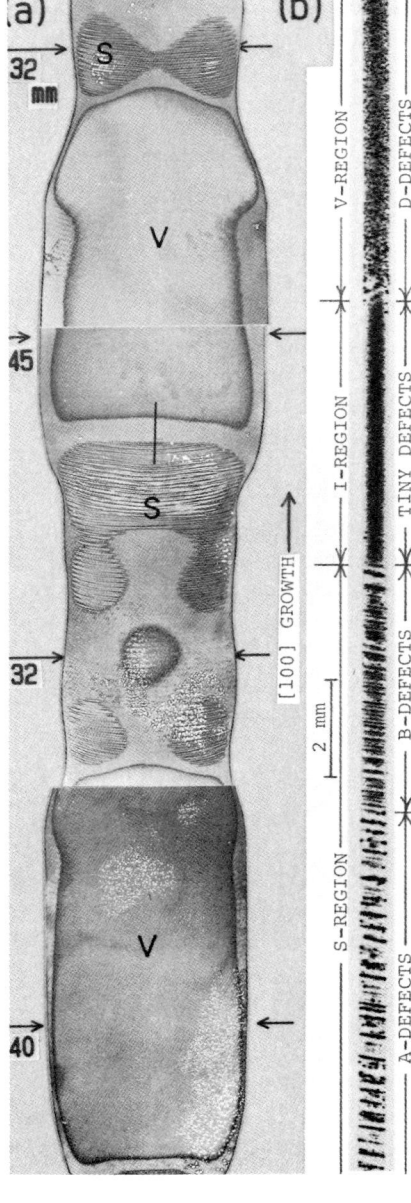

Fig. 1. Dependence of microdefect formation on crystal diameters. During FZ growth, the diameter was changed by adjusting the RF input power. (a) Topograph of Cu-decorated longitudinal specimen. S: Regions with swirl defects. V: Regions with D-defects. (b) Section topograph of the Cu-decorated specimen taken at the position of the vertical line at the center of (a). This photograph is negative, whereas (a) is a positive one.

of $T \gtrsim 1000°C$ and $C_V^{eq} > C_I^{eq}$ for $T \leq 1000°C$; for small ξ they have $C_I > C_I^{eq}$ and, consequently, $C_V < C_V^{eq}$ from eq. (1), and when ξ is large, eq. (1) does not hold, i.e., $C_I > C_I^{eq}$ and $C_V > C_V^{eq}$, and vacancies at larger concentration are frozen to form D-defects.

The present author, however, proposed a different model where vacancies play a principal role in the entire range of temperature.

3. Impurity and Microdefects

Miyamoto et al.[16] reported that swirl defects were not generated in FZ crystals grown after float-zoning several times in high vacuum; when such crystals were melted again and grown by introducing a small amount of oxygen into the atmosphere, swirl defects were generated. This result suggests that swirl defects (A and B defects) are not due to self-interstitials in equilibrium near the melting point but formed by effect of some kind of impurity such as oxygen.

The present author and his coworkers[11] found a new type of microdefects in FZ crystals free from both swirl and D-defects, employing X-ray topography following copper decoration. An example of such observations is shown in Fig. 1. By changing the diameter of a growing crystal, we can generate A, B, and D defects; Figure 1(a) is an X-ray topograph of a longitudinal specimen cut from such a crystal which was taken after copper decoration. The growth rate was fixed at 4 mm/min for the entire of the growth. Swirl defects are formed in the regions "S" by thermal fluctuation due to both increasing and decreasing the diameter. D-defects appear in the regions "V" with diameters of 40 and 45 mm and are not formed in the regions having the smaller diameter. (Regions with D-defects are referred to as "V-regions"). Figure 1(b) is a section topograph for the Cu-decorated specimen which was taken at the position indicated by the vertical line in Fig. 1(a). In the V-region, many black spots (D-defects) are seen in uniform distribution. In the S-region, A and B defects are seen along growth striations. Growth striations consists of the major and minor striations which are formed by remelting and temperature fluctuation at the interface, and A and B defects are seen along the major and minor striations, respectively. The region between V- and S-regions is seen black in the section topograph; it was found by enlargement that small black dots (Cu precipitates) distribute densely. Such regions are called "I-regions". The result observed on such a section topograph is illustrated schematically in Fig. 2(a); A and B defects are in striated distribution, and Pendellösung fringes are seen between the defect images. (Fringes appear for perfect crystals). Whereas, no Pendellösung fringes are seen in I-regions. Some of the Cu precipitates are a little larger and are in striated distribution (indicated by the arrows). It was found by SIMS measurement that Cu concentration in this I-region is in the same level as

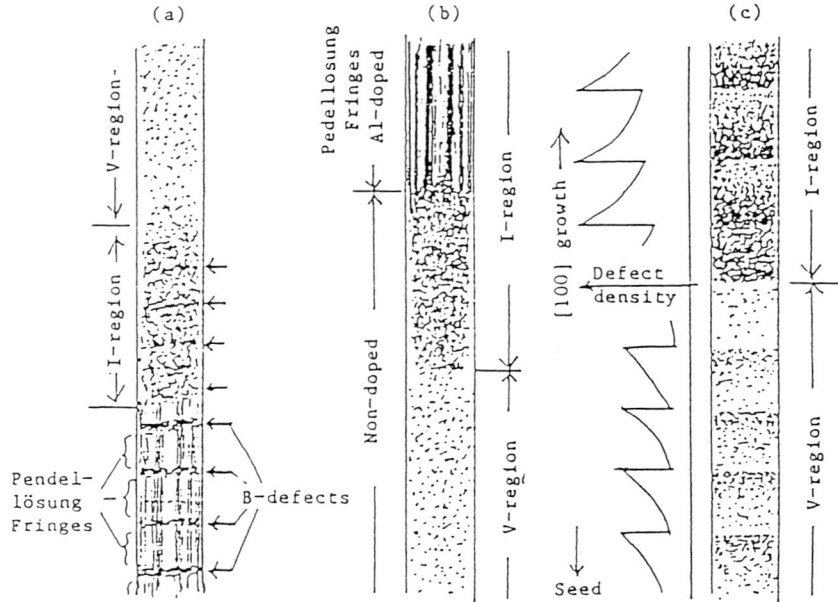

Fig. 2. Schematic illustration of X-ray section topographs taken after copper-decoration. (a) Non-doped crystal. (b) Al- or Ga-doped crystal. (c) Sb-doped. The variation in density of D-defects agrees with that in Sb concentration. Such striated distribution was observed for Bi- and In-doped crystals.

S-regions and an order of magnitude higher than that in V-regions. This suggests that defects in I-regions and S-regions are essentially similar, but the defects are distributed uniformly in I-regions and in striated patterns for S-regions, i.e., the defects in I-regions are considered to be of interstitial type.

The characters of defects in I-regions (named "I-defects") are enumerated:

(1) I-defects always appear after V-regions recede by annealing at high temperatures.

(2) No I-defects are formed by doping Al or Ga [Fig. 2(b)]. The other dopants, B, In, P, Sb, Bi, do not show such effect on defect formation, i.e., I-defects are generated.

(3) By doping impurities having large covalent radii, I-defects are in striated distribution with the opposite phase to the pattern of D-defects [Fig. 2(c)]. The striated distribution of D defects indicates that vacancies are attracted by expansive stress fields.

(4) By doping oxygen, densities of I-defects are increased. When doped oxygen is in striated distribution, I-defects are distributed similarly in striation.

It is concluded from these results that I-defects are microprecipitates of oxygen. It was reported that precipitates of Al_2O_3 are formed in CZ crystals doped with Al.[17] When two substitutional Al or Ga atoms form Al_2O_3 or Ga_2O_3 by taking three oxygen atoms, the resulting volume expansion is very small [Volume per Si atoms in the Si lattice = 20 A^3. Volume for $1/2(Al_2O_3) = 21$ A^3. Volume for $1/2(Ga_2O_3) = 24$ A^3. Compare with a molecule of cristobalite $SiO_2 = 43$ A^3]. Consequently, microprecipitates in Al- or Ga-doped crystals cannot be observed by Cu-decoration.

FZ crystals contain residual oxygen impurity in a concentration level of 10^{16} cm^{-3}. Solubilities of oxygen in Si were measured by many investigators. The values obtained are largely different in low temperatures, but converge at the melting point as seen in Fig. 3. This indicates that their measurements are reliable and solubilities vary depending upon quality of specimen crystals. Since silicon apparently tolerates very high oxygen supersaturation, the intrinsic solubilities are expected to be much lower than the measured values. Therefore, if we take the lowest value in Fig. 3 (the data by Logan and Peters[23]), oxygen concentration as low as 10^{16} cm^{-3} still exceeds the solubility at 1000°C.

By doping oxygen into FZ crystals, both oxygen and D-defects are distributed in striated patterns, similarly to the case of doping Sb [See Fig. 2(c). Oxygen impurity acts to expand the silicon lattice]. Such observation

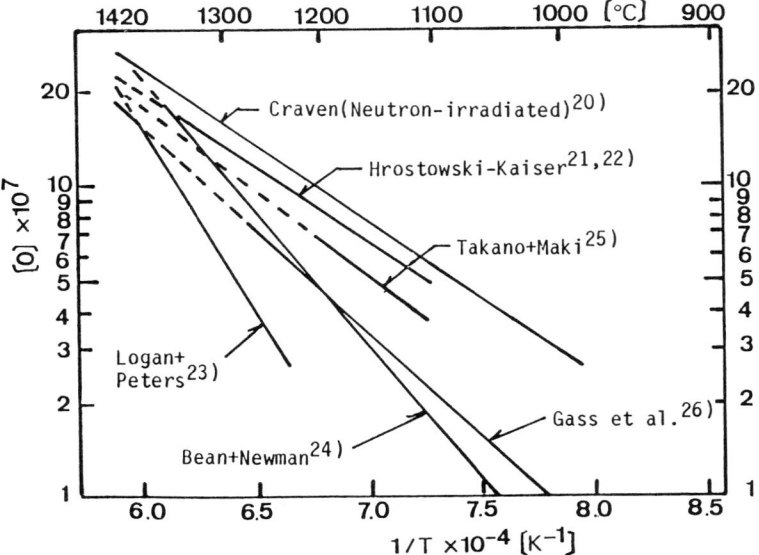

Fig. 3. Solubility of interstitial oxygen in silicon.

also showed that vacancy diffusion in oxygen-doped crystals is slower than that in non-doped crystals, i.e., oxygen atoms and vacancies interact elastically with each other. Therefore, interstitial oxygen atoms are stabilized by supersaturated vacancies; vacancies relax the strain energy of the silicon lattice due to interstitial oxygen atoms which is the driving force for precipitation, i.e., oxygen solubility is increased by supersaturated vacancies. However, tolerance (incubation time) for oxygen precipitation is decreased.

The vacancy concentration in the cooling process of crystal growth depends upon the growth condition, i.e., the growth rate v and temperature gradient G_O at the interface. For a given growth rate, the cooling rate is nearly constant from the melting point to 900°C and greatly slows down in the lower temperature region. In the case where high concentrations of vacancies at high temperatures hold until the crystal is cooled down to the lower temperature region, oxygen precipitation does not take place in the higher temperature region, and D-defects are formed around 600°C (experimental result for 42-mm crystal diameter). If out-diffusion of supersaturated vacancies occurs in a region of 800 to 900°C, then the oxygen concentration greatly exceeds the intrinsic solubilities at low temperatures (≤ 750°C), and oxygen microprecipitates are formed homogeneously. They are I-defects. When vacancy out-diffusion takes place at high temperatures, the oxygen concentration exceeds slightly the intrinsic solubilities around 1000°C, and precipitates are formed at active nucleation centers along the major growth striation due to remelting during growth. A-defects, the extrinsic dislocation-loops are formed by precipitation-induced self-interstitials. If the vacancies diffuse out

Table 2. Microdefects observed in float-zone crystals by X-ray topography following copper-decoration.

A-defect: Interstitial-type dislocation-loops (TEM)
    $T \sim 950$°C   Striated distribution

B-defect: Inferred as interstitial-type (Invisible by TEM)
    $T \sim 900$°C   Striated distribution

D-defect: Vacancy agglomerates (Invisible by TEM)
    $T \sim 600$°C   Uniform distribution
    Never exsist with A or B-defects
    Striated distribution in Sb, Bi, In, or O-doped crystals

I-defect: Very small defects in regions free from A, B, and D-defects
    Mainly $T \leq 750$°C   Uniform distribution
    Invisible in Al or Ga-doped crystals

(T: Formation temperature)

at a lower temperature region around 900°C, precipitation takes place at less active nucleation centers along minor striations (B-defects). The formation temperatures of A-, B-, and I-defects are observed to become higher by increasing oxygen concentration.

The formation of SiO_2 is supported with some experimental results: Ravi and Varker[18] observed grains of cristobalite inside swirl defects in FZ crystals, employing the TEM technique. Also, A- and B-defects are not formed in FZ crystals grown in 90%-argon/10%-hydrogen atmosphere (12); hydrogen may reduce oxygen concentration in the silicon melt.

The nucleation centers are considered to be formed near the growth interfaces. It was observed by *in-situ* X-ray observation of melting and solidification of silicon ribbons that, during melting of dislocation-free Si crystals, locally molten regions (drops) are always formed inside the crystals.[14] In conventional crystal growth such as the FZ and CZ methods, the temperature fluctuation during growth occurs due to the crystal rotation and melt convection; the fluctuation often reaches 10°C. Therefore, crystals grow by repeating growth and pausing (or remelting). In the pausing periods, the interface region is superheated highly enough to form drops. In the regrowth period, the drops solidify. Most of impurity atoms in Si have segregation coefficients less than unity, and, after solidification of the drops, they segregate at their central regions which is solidified finally. Such regions may act as nucleation centers.

4. Conclusion

Two kinds of microdefects, swirls and D-defects, have been observed in the conventional FZ crystals. In addition, a new type of very small defects named "I-defects" was recently found to exist in swirl- and D-defect-free regions. Thus, all dislocation-free FZ crystals contain, at least, one of these kinds of microdefects.

It is concluded that vacancies play a principal role on their formation and interactions between vacancies and impurities are very important; oxygen solubilities are enhanced by increasing concentration of supersaturated vacancies. Consequently, vacancy-controlled precipitation of residual oxygen impurity is responsible for formation of A-, B-, and I-defects: One of these types of defects is formed by oxygen impurity remaining in FZ crystals, depending upon the temperature that out-diffusion of vacancies takes place at.

Since vacancies are the dominant intrinsic defect of silicon, the vacancy concentration in a growing crystal is given as a function of v/G_O by the equation with omitting the third term for recombination with interstitials in eq. (4). Although the radial out-diffusion is not considered in eq. (4), the crystal diameter is also an important parameter to determine the vacancy concentration. Experimental results for diameters larger than 40 mm showed that D-defects are formed for $v/G_O > 3 \times 10^{-5}$ cm^2 K^{-1} s^{-1}.

Czochralski-grown (CZ) crystals have been used for fabrication of integrated circuits, because of a high mechanical strength due to oxygen impurity. In conventional CZ growth, the growth rate v and temperature gradient G_O are about 1/2 and 1/3 of those of FZ growth, respectively. Therefore, high vacancy concentrations hold in the low temperature region so that swirl defects are not formed in the as-grown state. Their oxygen concentration is 5 to 10×10^{17} cm^{-3}. Consequently, sizes and densities of oxygen microprecipitates are much larger compared with those of FZ crystals. Such defects in the active area of MOS transistors or capacitors cause deterioration in characteristics of highly integrated circuits such as 256-kbit dynamic random access memory (DRAM). Even smaller defects will become a serious problem for a new generation of electronic circuits with a high degree of integration and a large area of silicon.

From this point of view, nitrogen-doped FZ crystals have been proposed as a candidate material for ULSI: FZ crystals with a concentration as low as 10^{15} cm^{-3} have a much higher yield strength than usual FZ crystals, similarly to CZ crystals.[19] Furthermore, formation of swirls and D-defects is prevented by doping nitrogen. In fact, MOS diodes fabricated with N-doped FZ wafers show excellent characteristics in breakdown voltage, compared with CZ, MCZ (magnetic-field CZ), and epitaxial wafers.

N-doped FZ crystals, however, still contain microprecipitates (I-defects) like usual FZ crystals. To remove residual oxygen impurity, it is necessary to carry out float-zoning several times in high vacuum.[16] This results in a high production cost. For further improvement, therefore, gettering will become an important technique.

REFERENCES

1) For instance, see *Defects in Silicon*, ed. W. M. Bullis and L. C. Kimerling (Electrochemical Society, Pennington, N. J., 1983).
2) V. V. Voronkov: J. Cryst. growth **59** (1982) 625.
3) A. Seeger, H. Foll, and W. Frank: *Radiation Damage and Defects in Semiconductors 1976*, ed. N. B. Urli and J. W. Corbett (Institute of Physics, Bristol and London, 1977), Inst. Phys. Conf. Ser. No. 31, p. 12.
4) K. Wada, N. Inoue, and J. Osaka: *Defects in Semiconductors*, ed. J. W. Corbett and S. Mahajan (Mat. Res. Soc. Symp. Vol. 14) (Elsevier, N. Y., 1983) p. 125.
5) M. Yoshida and K. Saito: Jpn. J. Appl. Phys. **6** (1967) 573.
6) B. J. Masters and E. F. Gorey: J. Appl. Phys. **49** (1978) 2717.
7) For instance, see J. W. Corbett, J. P. Karins, and T. Y. Tan: Neclear Instruments and Methods **182/183** (1981) 457.
8) For instance, see M. Yoshida: Oyo Buturi **50** (1981) 2 (in Japanese); Also, see A. Seeger and K. P. Chik: Phys. State. Sol. **29** (1968) 455.
9) T. Y. Tan, F. Morehead, and U. Gosele: in Ref. (1), p. 325.
10) R. J. Roksnoer and M. M. B. Van den Boom: J. Cryst. Growth **53** (1981) 563.

11) T. Abe, H. Harada, and J. Chikawa: in Ref. (4), p. 1.
12) H. Föll and B. O. Kolbesen: Appl. Phys. **8** (1975) 319.
13) P. M. Petroff and A. J. R. de Kock: J. Crystal Growth **35** (1976) 4.
14) J. Chikawa and S. Shirai: Jpn. J. Appl. Phys. **18** (1979) Suppl. 18-1, 153.
15) S. Mizuo and H. Higuchi: Jpn. J. Appl. Phys. **20** (1981) 739.
16) N. Miyamoto, T. Sakai, and J. Nishizawa: *Proc. Gakujutsushinkokai No. 145 Comm. Meet.* (1978) p. 49.
17) R. Bullough, R. C. Newman, J. Wakefield, and J. B. Wills: J. Appl. Phys. **31** (1960) 707.
18) K. V. Ravi and C. J. Varker: *Semiconductor Silicon 1973*, ed. H. R. Huff and R. R. Burgess (Electrochem. Society, Princeton, 1973) p. 670.
19) T. Abe, K. Kikuchi, S. Shirai, and S. Muraoka: *Semiconductor Silicon 1981*, ed. H. R. Huff, R. J. Kriegler, and Y. Takeishi (Electrochemical Society, Pennington, N. J., 1981) p. 54.
20) R. A. Craven, in Ref. (19), p. 254.
21) H. J. Hrostowski and R. H. Kaiser: Phys. Rev. **107** (1957) 966.
22) H. J. Hrostowski and R. H. Kaiser: J. Phys. Chem. Solids **9** (1959) 214.
23) R. A. Logan and A. J. Peters: J. Appl. Phys. **30** (1959) 1627.
24) A. R. Bean and R. C. Newman: J. Phys. Chem. Solids **33** (1972) 255.
25) Y. Takano and M. Maki: in Ref. (18), p. 467.
26) J. Gass, H. H. Muller, and H. Stuss: J. Appl. Phys. **51** (1980) 2030.

Defects and Properties of Semiconductors: Defect Engineering, edited by J. Chikawa, K. Sumino, and K. Wada, pp. 155–167.
© KTK Scientific Publishers, Tokyo, 1987.

THE BEHAVIOR OF POINT DEFECTS IN SILICON CRYSTALS

Shoichi MIZUO and Hisayuki HIGUCHI

Central Research Laboratory, Hitachi Ltd., 1-280 Higashikoigakubo, Kokubunji, Tokyo 185, Japan

Abstract. The recent results of experiments involving point defects (interstitials and vacancies) have been summarized. The results can be explained by a model based on the thermal equilibrium between interstitials and vacancies.
  The effects of surface films and heat treatment ambients on point defect concentrations are discussed. Moreover, the importance of the role of atomic number in the crystal lattice has been discussed.

1. Introduction

Recently, process simulation has been widely used to predict impurity profiles for suppression of LSI development turnaround time.[1] The accuracy of the prediction is dependent on the diffusion models used in the simulation.
  Formerly, impurity diffusion in Si has been explained by vacancy based models.[2] However, the results of recent research have clarified the problems of the former models and the fact that the existence of interstitials has to be taken into account.[3]
  This paper will summarize recent research results. Most of these results are concerned with the diffusion of impurity atoms in Si under special boundary conditions such as oxidizing Si–SiO_2 interface, nitridating Si–SiO_2 and Si–Si_3N_4 interfaces, and using Si–SiO_2 and Si–Si_3N_4 interfaces. These results enable us to draw up consistent point defect models in Si. Moreover, remaining problems and suggestions for further investigation will be given.

2. Point Defects in Si

Formerly, vacancies were regarded as the dominant point defects in Si.[2] However, coexistence of oxidation enhanced diffusion (OED) of B and P, and oxidation induced stacking faults (OSFs) during thermal oxidation necessitates taking the presence of interstitials into account.[4-6]

It has been considered that the diffusion of group III and V elements proceeds via point defects in Si crystals.[7] Therefore, diffusion coefficients of these elements should be proportional to the concentration of point defects such as interstitials and vacancies.

Occurance of OED shows an increase in the concentration of point defects responsible for impurity diffusion during oxidation. Growth of OSF (extrinsic stacking faults) results in an increase in atomic concentration in the crystal. That is, an increase in interstitial concentration or a decrease in vacancy concentration.

Coexistence of both OED and growth of OSFs can only be explained by impurity diffusion via interstitials.[8] Thus, difficulties were encountered with impurity diffusion models based only on vacancies.

More recently, the authors have found oxidation retarded diffusion (ORD) of Sb in Si.[9] This corresponds to a decrease in concentration of point defects contributing to Sb diffusion during oxidation.

Therefore, coexistence of both OED and ORD of impurities means that at least two different kinds of point defects must exist in Si crystals. This situation can only be explained by assuming that there is a thermal equilibrium between interstitials, I, and vacancies, V, as shown below.

$$I + V = 0 . \qquad (1)$$

Thermal equilibrium of eq. (1) predicts the relation of mass action law.[9]

$$C_I \times C_V = \text{constant} , \qquad (2)$$

where C_I and C_V denote the concentration of interstitials and vacancies, respectively. The product of concentration of both interstitials and vacancies is constant. That is, an increase in the concentration of one type of point defect necessarily causes a decrease in the concentration of other type. Therefore, coexistence of both OED and ORD can well be explained. That is, oxidation of Si inject extra Si atoms into the crystal and causes both interstitial super-saturation and vacancy under-saturation.

3. Impurity Diffusion Mechanisms in Si

It was explained in the previous section that oxidation of Si causes interstitial super-saturation and vacancy under-saturation. Therefore, impurities diffusing by interstitials tend to diffuse faster during oxidation and those diffusing by vacancies tend to diffuse slower. Impurity diffusion coefficient via interstitials and vacancies can generally be expressed by the following equation.[10]

$$D_A = D_{AI} + D_{AV} = C_I d_{AI} + C_V d_{AV}, \qquad (3)$$

where D_A denotes diffusion coefficient of the impurity; D_{AI} and D_{AV} denote impurity diffusion coefficient via interstitials and vacancies. C_I and C_V denote concentrations of interstitials and vacancies, and d_{AI} and d_{AV} denote impurity diffusivity for unit concentration of point defects.

Tan et al. has calculated the effect of point defect concentration change on impurity diffusion under thermal equilibrium of interstitials and vacancies. This effect is graphed in Fig. 1.[11] The vertical axis is the relative impurity diffusivity and the horizontal axis is the change in interstitial concentration, s_I. The f_I value denotes the fraction of impurity diffusivity by interstitials (D_{AI}/D_A). For f_I greater than 0.5, an increase in interstitial concentration inevitably enhances diffusion. However, if $0 < f_I < 0.5$, it does not necessarily retard the diffusion. In this case, a small s_I tends to retard diffusion but a larger s_I tends to enhance diffusion.

The effect of oxidation on six impurity elements is summarized in Fig. 2.[12] The horizontal axis is the ionic radius for each impurity element. The vertical axis is the ratio of impurity diffusivity between oxidizing and non-oxidizing ambients.

Aluminum, boron, gallium and phosphorus show an equal extent of OED. On the other hand, Sb diffusion is retarded by oxidation. Arsenic shows a very weak diffusivity enhancement.

The authors consider that Al, B, Ga and P diffuse entirely via interstitials and that Sb diffuses entirely via vacancies. The only exception is As, which may diffuse via both interstitials and vacancies. The reasons for this are: The

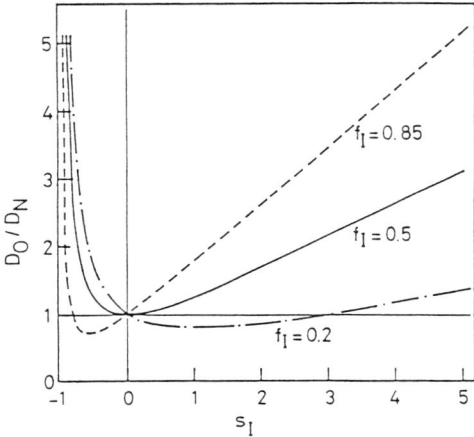

Fig. 1. Relation between the extent of interstitial super-saturation and the ratio of diffusivity in oxidizing and non-oxidizing ambient (after Tan and Gosele ref. 11).

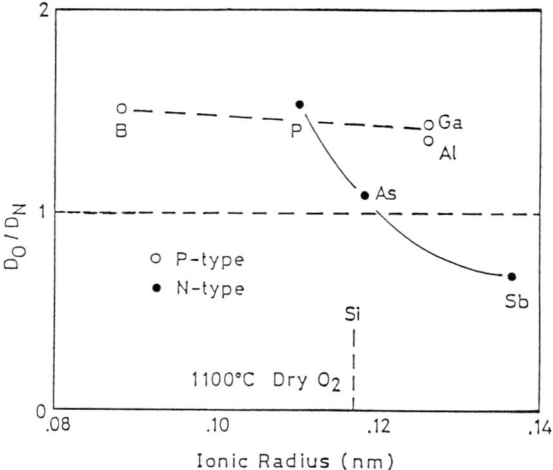

Fig. 2. Relation between ratio of diffusivity in oxidizing and non-oxidizing ambients and ionic radius for six impurity elements.

agreement of OED extent in the four elements, in spite of the difference in their ionic radii and periodic groups, and the similarity of the extent of both OED and ORD for Al, B, Ga and P, and Sb.

Other researchers have proposed that most impurity elements diffuse by a dual mechanism.[13,14] Antoniadis et al. proposed that f_I is 0.38 for P and 0.30 for B at 1000°C.[13] That is, they considered that s_I is very large and thus the two impurities show OED in spite of the small f_I value. They estimated s_I from the OSF growth data and calculated f_I using s_I. At present, it is not clear which model is valid, and further investigation is necessary. However, as will be shown later, B and P show retarded diffusion during Si nitridation at 1100°C and therefore f_I for these elements should be greater than 0.5 at this temperature.

4. Effects of Si and SiO₂ Nitridation on Impurity Diffusion

Oxidation of Si is believed to increase interstitial concentration and to decrease vacancy concentration. Effects of nitridation of both Si and SiO_2 on impurity diffusion and growth kinetics of OSFs have been investigated by Hayafuji et al., the authors and Fahey et al.[15-17] Results are summarized in Table 1.

Nitridation of Si has been found to cause retarded diffusion of B and P, and enhanced diffusion of Sb and rapid shrinkage of OSF. On the other hand, nitridation of SiO_2 on Si crystal causes enhanced diffusion of B and P, retarded Sb diffusion and growth of OSF.

Table 1. Effect of nitridation on impurity diffusion and OSF length.

	Oxide Nitridation	Silicon Nitridation
B, P Diffusion	Enhanced Diffusion	Retarded Diffusion
Sb Diffusion	Retarded Diffusion	Enhanced Diffusion
OSF Length	Growth	Shrinkage

These phenomena suggest that nitridation of Si extracts excess Si atoms from the crystal and causes interstitial under-saturation and vacancy super-saturation. Nitridation of SiO_2 on Si apparently injects excess Si atoms into the Si crystal and causes interstitial super-saturation and vacancy under-saturation. The mechanisms of these phenomena are not clear at present. However, they are probably closely related to the mechanism of Si and SiO_2 nitridation.

5. Interaction between Point Defects and Surface Films on Si Crystals

Oxide and nitride films formed on the Si have different characteristics for point defects in the crystal. A cross section of the sample and results of this experiment are shown in Figs. 3 and 4.[18,19] There are two areas on the sample surface: One is the area masked with a Si_3N_4 film (N-area) and the other is that masked with a double-layered SiO_2–Si_3N_4 film (ON-area). This sample was heat-treated at 1100°C and the junction depth, x_j, in both areas was measured and compared. x_j in the ON-area increases proportionally with the square-root of the oxidation time. This suggests that diffusion in the ON-area is normal. That is, the diffusion coefficient is independent of heat treatment

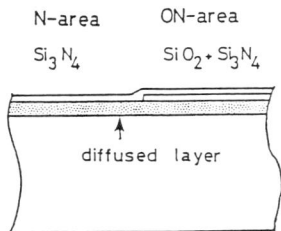

Fig. 3. Sample cross section for diffusion experiment under Si_3N_4 films.

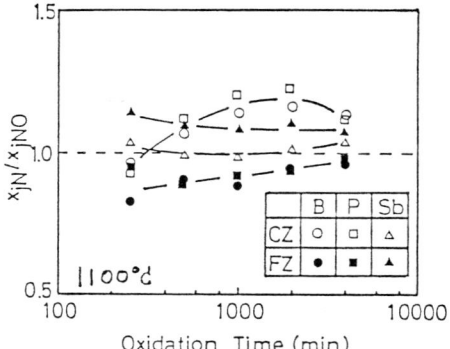

Fig. 4. Relation between normalized junction depth and heat treatment duration.

time. The ratio of x_j of both the N and ON-area is shown in Fig. 4 in relation to the heat treatment time.

Anomalous enhanced diffusion is found in the N-area for CZ substrates. This anomalous diffusion is suppressed by a pre-diffusion annealing, the so-called intrinsic gettering (IG) process.[20] Therefore, this anomalous diffusion enhancement can be regarded as being due to oxygen precipitation and subsequent interstitial super-saturation in the CZ Si. Antimony diffusion in CZ Si seems to have little interfacial effect. This may be due to the high Sb pre-deposition diffusion temperature (1175°C). Pre-deposition heat treatment will likely have the same effect as IG treatment.

Enhanced Sb diffusion and retarded B and P diffusion can be seen for FZ substrates. This may be caused by the mismatch between the atomic number and the number of lattice sites and will be discussed later.

6. Diffusion of Point Defects in Si Crystals

Point defects in Si not only contribute to impurity diffusion but also diffuse themselves. Backside selective oxidation (BSO) has been used to clarify the diffusion of point defects in Si.[21,22] A cross section of the sample used in this experiment is shown in Fig. 5(a).

There is an impurity-diffused layer with Si_3N_4 films at the front side of the sample. There are two areas at the back, one is covered with an oxidation resistant film (BN-area), and the other bare and etched to control the thickness in this area (BO-area). Samples were then heat-treated at 1100°C in dry O_2. The back surface of the BO-area is oxidized and the concentration change of point defects spreads towards the front side by diffusion. The change in point defect concentration at the front surface varies the diffusivity of impurities.

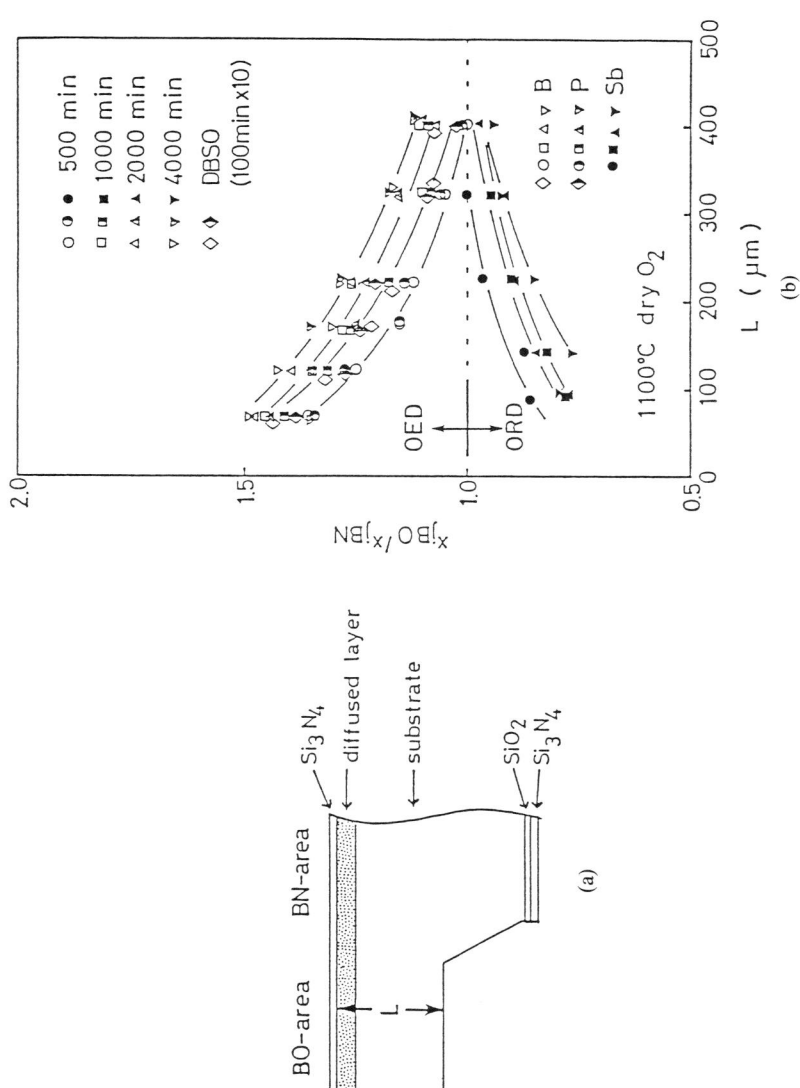

Fig. 5. Effect of backside oxidation on impurity diffusion: (a) sample cross section; (b) results.

Results of the experiment are shown in Fig. 5(b). Here, the horizontal axis is the thickness (L) of the samples in the BO-area and the vertical axis is the ratio of x_j in the BO and BN-area. Boron and phosphorus have ratios larger than unity, indicating enhanced diffusion.

On the other hand, Sb diffusion is retarded. The oxidation effect is significant with smaller L and longer oxidation time. Results show that excess Si atoms generated at the oxidizing Si–SiO$_2$ interface diffuse towards the front surface and change point defect concentration. Moreover, the strength of the oxidation effect, and L and oxidation time dependence are similar for B, P and Sb. This suggests the presence of thermal equilibrium between interstitials and vacancies.

The effect of backside oxidation on impurity diffusion at the frontside is found only in the case when a Si$_3$N$_4$ is directly formed at the front surface. Impurity diffusion under a double-layered SiO$_2$–Si$_3$N$_4$ film is not affected by backside oxidation. This supports the validity of the explanation that Si–SiO$_2$ and Si–Si$_3$N$_4$ interface have different interfacial characteristics. Moreover, it is noteworthy that effect of BSO is found only for FZ substrates. CZ substrates show neither OED nor ORD by BSO. Macroscopic defects are formed during heat treatment in CZ Si.[23] These defects may act as sinks for excess point defects. Therefore, BSO effect is found only in FZ substrates.

The apparent diffusivity of point defects calculated from the results are on the order of 10^{-9} cm^2/s[24] and agree with the value obtained from lateral point defect diffusion experiment.[25]

The diamonds denote results obtained after the samples were oxdized ten times (100 min each time) (Divided Backside Selective Oxidation, DBSO).[26] The DBSO results are close to those for 1000 min BSO. This corresponds to the preservation of point defect distribution after heat treatment.

If point defect distribution is not preserved after heat treatment, the distribution of point defects at the beginning of the next heat treatment is the same as that at the beginning of the prior heat treatment. Therefore, the distribution of point defects after heat treatment becomes shallower by dividing heat treatment. On the other hand, if the distribution is preserved after heat treatment, the distribution of excess point defects during the next heat treatment is the same as that at the end of the prior treatment. Therefore, the range of the excess point defect distribution is determined by the sum of the heat treatment time and is independent of each separate heat treatment time. Results of the DBSO experiment show that point defect distribution is determined by the sum of the heat treatment time and suggest that the point defect distribution is preserved.

7. Mismatches between the Atomic Number and Number of Lattice Sites in Si

The problem of non-stoichiometry for compound semiconductors such as GaAs or InP is well known. However, for elemental semiconductors such as Si and Ge this kind of problem does not occur. However, the x_j discrepancy between the ON and N-area (section 5) has made it clear that this type of problem does exist in elemental semiconductor crystals.[27]

Silicon has a diamond structure and the number of lattice sites (N_L) in the unit cell is 8. However, the number of atoms (N_A) actually present in the unit cell is not necessarily 8. If there were only interstitials in Si crystals, N_A would be greater than 8. On the other hand, if there were only vacancies, N_A would be smaller than 8. If both types of defects, N_A would be either greater or smaller than 8.

If we let the number of interstitials and vacancies in a unit cell be N_I and N_V, the following relation holds,

$$N_A - N_L = N_I - N_V = N_M , \tag{4}$$

where N_M denotes the mismatch between the atomic and lattice site numbers.

It is noteworthy that N_M does not change due to the equilibrium reaction. This is because the generation and recombination of interstitials and vacancies occur as a paired phenomenon. Change in N_M can only be possible through the reaction of individual point defects (interstitials or vacancies) and the crystal surface or macroscopic defects in the crystal (such as dislocations or OSFs). It has been shown that anomalous retarded B and P diffusion and Sb-enhanced diffusion occur in FZ substrates in the N-area. This is probably due to a mismatch between N_A and N_L which occurs during crystal growth. That is, an FZ Si crystal has a fixed value of N_A corresponding to the crystal growth condition ($N_{A, r}$). When the sample was heat treated at 1100°C, the $N_{A, r}$ value plays a dominant role in determining the point defect concentration. If $N_{A, r}$ is larger than the equilibrium value of N_A at 1100°C ($N_{A, eq}$) the excess Si atoms act to increase the interstitials and decrease the vacancies. Therefore, interstitial concentration (C_I) is larger than the equilibrium value ($C_{I, eq}$), and vacancy concentration (C_V) is smaller than the equilibrium value($C_{V, eq}$). If $N_{A, r}$ is smaller than $N_{A, eq}$, C_I would be smaller and C_V would be larger than the equilibrium values. Enhanced Sb diffusion and retarded B and P diffusion under directly formed Si_3N_4 films correspond to smaller interstitial concentration and larger vacancy concentration. Hence, $N_{A, r}$ would be smaller than $N_{A, eq}$.

8. Calculation Procedures for Precise Diffusion Simulation

Impurity diffusion coefficients are thought to be constant throughout Si crystals.[28] This is a safe assumption if there is an oxide film at the top of the crystal. However, in the case of local oxidation of Si or diffusion under directly formed Si_3N_4 films, this assumption is no longer valid. In these cases impurity diffusion coefficient distribution must be taken into account in relation to the distribution of point defects. The calculation procedures to be followed in such cases are shown in Fig. 6. Calculation step 1 involves point defect diffusion calculation, with which point defect distribution can be obtained. In step 2, diffusion of impurities is calculated by using local point defect concentration to calculate local impurity diffusion coefficients. These two steps are repeated corresponding to the time of heat treatment.

9. Remaining Problems

Recent research results have clarified the behavior of point defects qualitatively. In this section, remaining problems and suggestions for the future will be given.

9.1 Concentration of interstitials and vacancies and their diffusion rate

Recent studies have clarified the thermal equilibrium between interstitials and vacancies and several of their important characteristics. However,

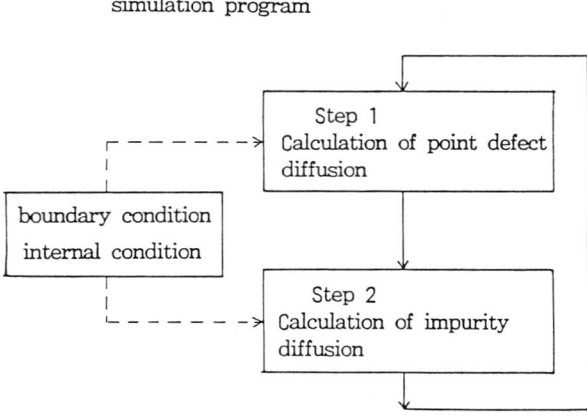

Fig. 6. Conceptual representation of calculation procedure flow for precise impurity diffusion simulation.

qualitative information about point defects is limited. It is necessary to obtain quantitative data for point defects, such as their concentrations (C_I and C_V), diffusion coefficients (D_I and D_V), and the rate of equilibrium reaction (r_{eq}). The experimental results for oxidation effect range on impurity diffusion will likely provide us with information of point defect diffusion.

However, it is noteworthy that the apparent diffusion coefficient in these experiment are not necessarily the actual diffusion coefficients. Diffusion flux of both interstitials and vacancies are given by $C_I D_I$ and $C_V D_V$. Diffusion of interstitials in one direction is equivalent to the diffusion of vacancies in the opposite direction. Therefore, diffusion of interstitials and vacancies are not independent of each other. The dominant diffusing point defects are those with larger CD value and not necessarily those with larger concentrations. If defects with larger concentration have larger CD value, the apparent diffusion coefficient obtained in the above experiment is the true coefficient for the major point defects. However, if not, the apparent coefficient is the diffusion coefficient of major diffusing point defects divided by the ratio of point defect concentrations.

It is necessary to know which are the major diffusing point defects. Towards this end, knowing the effect of oxidation or nitridation on Si self-diffusion would be helpful. If interstitials are the major diffusing point defects, self-diffusion will be enhanced by oxidation and retarded by nitridation. The reverse is true if vacancies are the major diffusing defects.

9.2 Effect of high concentration impurity diffusion on point defect equilibrium

In actual ICs and ISIs, impurity diffused layers with high concentration are usually applied. Hence, it is important to simulate high concentration diffusion accurately.

Formerly, investigations of high concentration diffusion have been carried out on the basis of sophisticated theories.[29] However, the basis for these theories (vacancy model) has proved incorrect.

Most of the recent works have been concerned with rather low concentration diffusion. Therefore, investigations of high concentration diffusion are necessary.

Harris et al.[30] and Fahey et al.[31] have investigated the effect of high concentration P diffusion on diffusion of buried diffused Sb and As. They found As diffusion was enhanced and Sb diffusion was retarded. Hence, they concluded that high concentration P diffusion causes super-saturation of interstitials and under-saturation of vacancies. Experiments of this type need to be carried out for other impurities.

9.3 Relation between atomic number in Si crystals and crystal growth condition

Diffusion under directly formed Si_3N_4 films in FZ Si have posed the problem of mismatches between atomic and lattice site number in Si. The experiment detailed in section 5 was carried out for commercially sold FZ substrates. The atomic number (N_A) is expected to be dependent on Si crystal growth conditions.

The relationship between crystal growth conditions and N_A is a remaining subject for investigation. Knowledge of this relationship would also be helpful in controlling and suppressing macroscopic defect generation (*e.g.*, OSF or dislocation generation), during semiconductor device processing.

9.4 Thermal equilibrium of point defects at lower temperatures

The presence of thermal equilibrium between interstitials and vacancies has been proven at high temperatures (above $1000°C$). The degree to which this equilibrium holds at lower temperatures has yet to be determined.

10. Conclusion

The behavior of point defects in Si was clarified by diffusion of impurities under various boundary conditions. The following results made it possible to draw a consistent view of the point defect structure in Si.

(1) There is a thermal equilibrium between interstitials and vacancies in high temperature Si.

(2) Mismatches between the atomic number and number of lattice sites plays a dominant role in determining the concentration of interstitials and vacancies.

(3) Oxidation of Si and nitridation of SiO_2 on Si lead to injection of excess Si atoms as well as causing interstitial super-saturation and vacancy under-saturation.

(4) Nitridation of Si leads to extraction of Si atoms from Si crystal and causes under-saturation of interstitials as well as super-saturation of vacancies.

(5) Antimony diffuses via vacancies, Al, B, Ga and P diffuse via interstitials, and As by both interstitials and vacancies.

(6) The $Si-SiO_2$ interface provides sinks and generation centers for interstitials and vacancies, but the $Si-Si_3N_4$ interface does not react with point defects in Si.

REFERENCES

1) R. W. Dutton and S. H. Hansen: Proc. of the IEEE **69** (1981) 1305.
2) R. B. Fair: J. Electrochem. Soc. **122** (1975) 800.

3) U. Gosele and H. Strunk: Appl. Phys. **20** (1979) 265.
4) M. Okamura: Jpn. J. Appl. Phys. **9** (1970) 848.
5) G. Masseti et al.: Solid State Electron. **16** (1973) 1419.
6) S. Prussin: J. Appl. Phys. **43** (1972) 2850.
7) S. M. Hu: *Atomic Diffusion in Semiconductors*, ed. D. Shaw (Plenum Press, London, 1973).
8) S. M. Hu: J. Appl. Phys. **45** (1974) 1567.
9) S. Mizuo and H. Higuchi: Jpn. J. Appl. Phys. **20** (1981) 739.
10) U. Gosele and T. Y. Tan: to be published in Proc. of MRS Symposium, *Impurity Diffusion and Gettering in Semiconductors* (Boston, Nov. 1984).
11) T. Y. Tan and U. Gosele: Appl. Phys. Lett. **40** (1982) 616.
12) S. Mizuo and H. Higuchi: Denki Kagaku **50** (1982) 338.
13) D. A. Antoniadis and I. Moskowitz: J. Appl. Phys. **53** (1982) 6788.
14) S. Matsumoto et al.: J. Appl. Phys. **54** (1983) 5049.
15) Y. Hayafuji et al.: J. Appl. Phys. **53** (1982) 8639.
16) S. Mizuo et al.: J. Appl. Phys. **54** (1983) 3860.
17) P. Fahey and R. W. Dutton: Appl. Phys. Lett. **43** (1983) 683.
18) S. Mizuo and H. Higuchi: Jpn. J. Appl. Phys. **21** (1982) 281.
19) S. Mizuo and H. Higuchi: Denki Kagaku **51** (1983) 403.
20) S. Mizuo and H. Higuchi: Jpn. J. Appl. Phys. **20** (1981) 1749.
21) S. Mizuo and H. Higuchi: J. Electrochem. Soc. **129** (1982) 2292.
22) S. Mizuo and H. Higuchi: J. Electrochem. Soc. **130** (1983) 1942.
23) W. Frank et al.: to be published in *Diffusion in Solid II* (A. S. Nowick and G. Murch Eds., Academic Press).
24) S. Mizuo and H. Higuchi: Jpn. J. Appl. Phys. **21** (1982) 272.
25) S. Mizuo and H. Higuchi: Jpn. J. Appl. Phys. **22** (1983) 12.
26) S. Mizuo and H. Higuchi: Jpn. J. Appl. Phys. **20** (1982) 1547.
27) S. Mizuo and H. Higuchi: to be published in Proc. of MRS Symposium, *Impurity Diffusion and Gettering in Semiconductors* (Boston, Nov. 1984).
28) D. A. Antoniadis et al.: Stanford Univ. Tech. Rep., SUSEL 77-066, May 1977.
29) R. B. Fair: *Impurity Doping Processes in Silicon*, ed. F. F. Y. Wang (North-Holand, Yew York, 1981) Chap. 7.
30) R. M. Harris et al.: Appl. Phys. Lett. **43** (1984) 937.
31) P. Fahey et al.: Appl. Phys. Lett. **44** (1984) 777.

POINT DEFECTS AND STACKING FAULT GROWTH IN SILICON

K. WADA and N. INOUE

NTT Electrical Communications Laboratories, 3-1, Morinosato Wakamiya, Atsugi, Kanagawa 243-01, Japan

Abstract. The growth of bulk stacking faults is experimentally- and theoretically studied in both bulk and surface region of Czochralski silicon by two-step annealing. First, the mechanism of nonequilibrium point defect formation and bulk stacking fault growth is studied. Next, a quantitative analysis of stacking fault growth in specimen bulk and surface is developed and experiments are performed. It is found that the temperature dependence of the bulk stacking fault growth rate is identical to that of the self-diffusion coefficient. Based on this result, a new model is proposed for the bulk stacking fault growth, where the dominant driving force of the growth is vacancies in undersaturation, not self-interstitials in supersaturation. Within the model, the vacancy component of the self-diffusion coefficient is obtained. This component is further divided into the diffusion coefficient and equilibrium concentration of vacancies. This development provides first step to elucidate the thermal properties of point defects in silicon.

1. Introduction

Vacancies and self-interstitials are the most fundamental defects in silicon and play paramount roles in device fabrication processes: They act as vehicles for group III and V impurity diffusion,[1] which is an elemental silicon device fabrication process. In addition, they drive the growth of dislocation loops (stacking faults) which act as the intrinsic gettering sites.[2] To precisely

Notation: r, radius of bulk stacking faults (cm); t, annealing time (s); v, volume of point defects, 2×10^{-23} (cm^3), given by C_S^{-1} assuming that vacancy is the same as self-interstitials in volume; C_S, lattice concentration, 5×10^{22} (cm^{-3}; D_S, self-diffusion coefficient; b, Burgers vector of bounding partial, 3.14×10^{-8} (cm); r_c, core radius (cm); C, point defect concentration (atoms/cm^3); D, point defect diffusion coefficient (cm^2/s); γ, stacking fault energy, 0.026 (eV/atom); g, growth contribution factor lying between 0 and 1; f, correlation factor; k, Boltzman constant, 8.265×10^{-5} (eV/K); T, annealing temperature (K); CD, tor dominating the temperature dependence of the stacking fault growth rate, given by eq. (2); z, depth under the specimen surface. Subscripts: V, vacancies; I, self-interstitials; E, equilibrium; ', values on the dislocation loop.

control group III and V impurity diffusion and stacking fault introduction, the thermal properties of these point defects should be clarified. However, knowledge on their thermal properties, such as diffusion coefficient and equilibrium concentration, is quite poor. Studies on the thermal properties obtained by diffusion studies are summarized in a preceding paper.[3] The present paper reviews recent studies on bulk stacking fault growth kinetics performed in our group of NTT Laboratories.

Growth of oxidation-induced surface stacking faults has been studied in conjunction with oxidation-enhanced (retarded) dopant diffusion.[4] This study is very important because the process is actually used in the device fabrication process. However, analysis of oxidation-induced surface stacking fault growth is highly complicated and the results regarding thermal properties are rather poor, despite of tremendous amount of publications.[4,5] In contrast, analysis of the bulk stacking fault growth combined with oxygen precipitation, has great advantages over the oxidation-induced surface stacking faults growth analysis. This is because the growth occurs in a closed system, as has been demonstrated in several publications.[6] As a result, the present paper limits its scope to the experimental and theoretical study of bulk stacking fault growth kinetics analysis. It treats bulk stacking fault growth in the specimen bulk and in the surface layer of bulk specimens. The following information is successfully extracted: The dominant driving force of bulk stacking fault growth is vacancies in undersaturation. In addition, the diffusion coefficient and equilibrium concentration are obtained.

2. Mechanism of Nonequilibrium Point Defect Formation and Bulk Stacking Fault Growth

2.1 Located in specimen bulk

This section describes the stacking fault growth process in the specimen bulk for a two-step annealing used in the present paper. Figure 1 illustrates the formation processes of nonequilibrium point defects and microdefects during the two-step annealing. During a lengthy first annealing near 800°C, oxide precipitates in the specimen bulk grow without nucleation of visible secondary defects (see (a)).[7] They consume oxygen interstitials (e) until achieving its equilibrium concentration ((b) and (f)). To accommodate volume expansion of the oxide precipitates, vacancies in undersaturation ((m) and (n)) and/or self-interstitials in supersaturation ((i) and (j)) must be concurrently induced in the specimens.

The stacking fault growth is driven by vacancies in undersaturation (Vacancy model) and/or self-interstitials in supersaturation (Self-interstitial model), which have been generated in the first step annealing. Therefore, these nonequilibrium point defects induce stacking faults at the oxide precipitates (c) during the high temperature second annealing around 1000°C.[8] In other

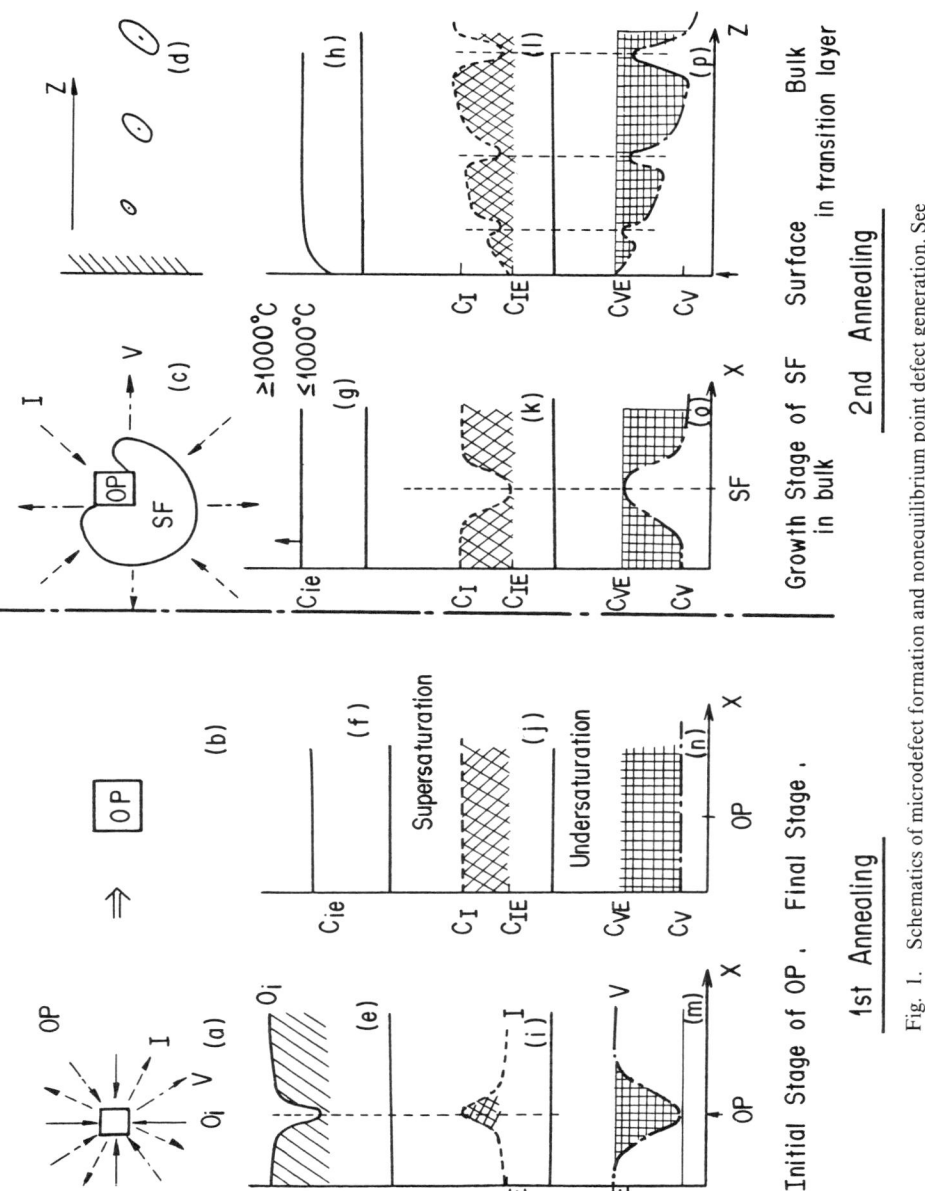

Fig. 1. Schematics of microdefect formation and nonequilibrium point defect generation. See Session 2.

words, vacancies are generated and/or self-interstitials are agglomerated at the bounding partials, respectively. This generation (source action) and agglomeration (sink action) decreases the degree of the point defect nonequilibrium condition ((h) and (o)). Thus, the bulk stacking faults keep growing until the point defects are in equilibrium with the bounding partials.

During the second annealing (higher than 1100°C), oxide precipitate redissolution occurs due to an oxygen solubility increase (g),[9] which reduces nonequilibrium point defects. This disrupts simplicity of the growth kinetics analysis, as will be shown later.

In the present paper, the stacking fault growth is analyzed by vacancy model and self-interstitial model.

2.2 Located in the specimen surface layer

The most significant characteristic of stacking fault growth in the specimen surface layer is that the surface acts as another sink or source in the vicinity of the specimen surface (transition layer). In this way, vacancies can in-diffuse toward the inside of specimens to decrease the degree of undersaturation (p). In addition, self-interstitials out-diffuse toward the specimen surface to decrease the degree of supersaturation (l). In-diffusion of vacancies and/or out-diffusion of self-interstitials results in growth retardation in the transition layer or depth profile formation (d). Thus, the indepth profile can be expressed by a simple error function, as shown in Section 4. Therefore, the diffusion coefficient of the point defects responsible for the stacking fault growth can be obtained.[10]

The difference in growth kinetics analysis between the bulk stacking faults in the transition layer and oxidation induced stacking faults in the surface layer is that nonequilibrium point defect concentration is changing during oxidation process and the concentration dependence on annealing time can not be expressed by a simple function. In this way, analysis of the oxidation-induced surface stacking fault growth is not so easy.

3. Growth of the Stacking Faults in the Specimen Bulk

3.1 Growth kinetic equation

First the following assumption was made: *The vacancies and self-interstitials in the specimen bulk reach their concentration minima and maxima, respectively, by oxygen precipitation in the low temperature first annealing ((j) and (n)). These conditions are maintained throughout the high temperature second annealing.* The growth rate of the bulk stacking faults is then expressed by:[6]

$$dr/dt = 2\pi v/b \ (1 - (r_c/r)^2)^{1/2}(\ln(8r/r_c))^{-1}CD , \qquad (1)$$

$$CD = g_V (C_{VE}' - C_V)D_V + g_I(C_I - C_{IE}')D_I$$

$$- (g_V C_{VE} D_V + g_I C_{IE} D_I) \gamma/kT. \qquad (2)$$

The first and second terms of CD in eq. (2) denote the growth of stacking faults driven by vacancy undersaturation and self-interstitial supersaturation, while the third term denotes stacking fault shrinkage which becomes dominant in inert atmosphere. The third term can be ignored in the stacking fault growth condition. Thus, eq. (2) can be rewritten as:

$$CD = g_V (C_{VE}' - C_V)D_V + g_I (C_I - C_{IE}')D_I. \qquad (2')$$

Therefore, the final expression for the stacking fault growth rate dependence on annealing temperature is obtained. This expression will be utilized in the next section.

3.2 Experimental procedures

Specimens used in this section were sliced from undoped, as-grown Czochralski silicon ingots of 3 inches in diameter. They were not exposed to heat treatment after growth. Oxygen and carbon concentrations were determined by the Infrared absorption measurements to be 1.1×10^{18} atoms/cm^3 and less than 2×10^{16} atoms/cm^3, respectively. The proportionality factors are 3×10^{17} for oxygen and 8×10^{16} for carbon. The specimens were two-step annealed in an argon atomosphere. The first annealing was at 750°C for 161 hours. The second annealings were performed between 850° and 1200°C. The stacking fault radius was measured using JEOL TEM 200C operated at 200 kV.

3.3 Results and discussion

Figure 2 shows the annealing time dependence of bulk stacking fault radius by the 2-step annealing.[6,11] The typical growth stage of bulk stacking faults is explained using data at 1000°C. Bulk stacking faults grow rapidly up to about 500 Å (r_0) in radius where the growth abruptly halts (growth halt stage). At a later stage, the growth begins again (steady growth stage). Finally, the growth saturates at a radius of about 6000 Å (growth saturation stage), showing that nonequilibrium point defects becomes in equilibrium with dislocation loops. Similar features are observed in this wide temperature range.

In this section, the steady growth stage is analyzed. The solid lines are obtained by curve fitting with eqs. (1) and (2), considering the growth halt and saturation. The fitting parameter is CD which is shown in Fig. 3. Thus, annealing temperature dependence of the stacking fault growth rate CD is given by:

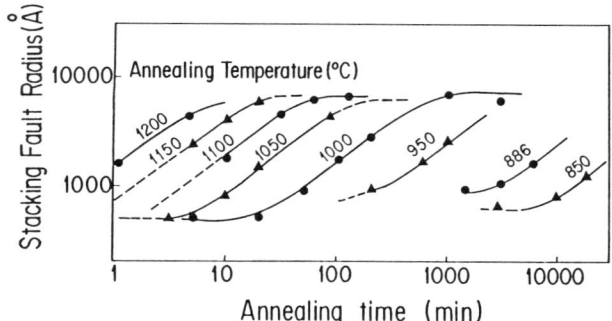

Fig. 2. Bulk stacking fault growth process. Parameter is annealing temperature (°C).

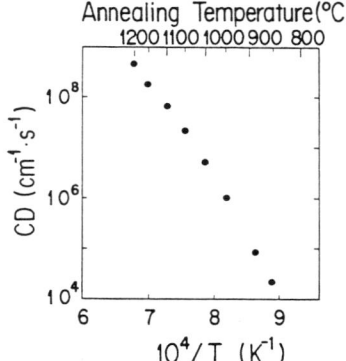

Fig. 3. Temperature dependence of CD in eq. (2). CD governs the bulk stacking fault growth rate.

$$CD = 2.2 \times 10^{24} \exp(-4.4/kT) \ (\text{cm}^{-1}\text{s}^{-1}) \ . \tag{3}$$

Here, data higher than 1100°C are ignored, since the nonequilibrium point defect concentration is probably changed by the oxide precipitate redissolution in the etemperature range.[9]

The following discussion aims at elucidating the point defects responsible for the stacking fault growth.

Figure 4 shows the temperature dependence of CD divided by lattice concentration C_S and the tracer self-diffusion coefficients[12-18] and self-diffusion coefficients obtained from the stacking fault shrinkage experiments.[19,20] A significant result is obtained:

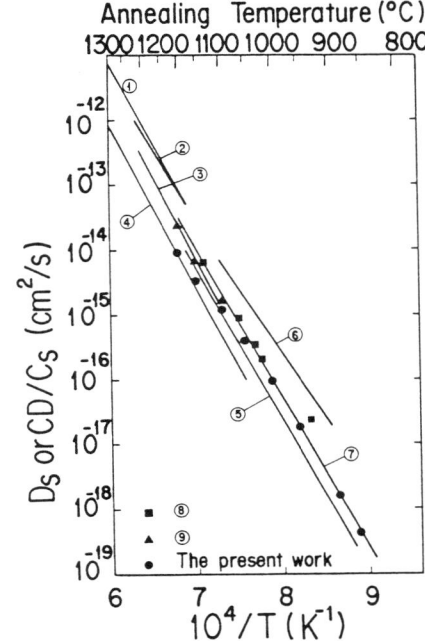

Fig. 4. Comparison between CD/C_S and D_S. CD/C_S is nearly equal to the tracer self-diffusion coefficient and the self-diffusion coefficient calculated from stacking fault shrinkage kinetics. 1: ref. 12, 2: ref. 13, 3: ref. 14, 4: ref. 15, 5: ref. 16, 6: ref. 17, 7: ref. 18, 8: ref. 19, 9: ref. 20.

$$CD/C_S = D_S. \quad (4)$$

This means that the temperature dependence of stacking fault growth rate follows that of self-diffusion. In this section, the relation between CD/C_S and the self-diffusion coefficients obtained from the stacking fault shrinkage experiments is discussed. The relation to the tracer self-diffusion coefficients will be discussed in the Section 5. This result helps in clarifying the relationship between point defects and stacking fault growth (shrinkage).

Using eq. (2'), the left hand side of eq. (4) is written as:

$$CD/C_S = (g_V(C_{VE}' - C_V)D_V + g_I(C_I - C_{IE}')D_I)/C_S \quad (5)$$

$$= (g_V(1 - C_V/C_{VE})C_{VE}D_V + g_I(C_I/C_{IE} - 1)C_{IE}D_I)/C_S.$$

Equation (5') is derived using the simplifications $C_{VE} = C_{VE}'$ and $C_{IE} = C_{IE}'$. On the other hand, the right hand side of eq. (4) is written as:

$$D_S = (g_V C_{VE} D_V + g_I C_{IE} D_I)/C_S . \tag{6}$$

Equating eq. (5′) to eq. (6), the following equations should be valid:

$$1 - C_V/C_{VE} = 1, \text{ or } C_{VE} \gg C_V , \tag{7a}$$

$$C_I/C_{IE} - 1 = 1, \text{ or } C_I = 2C_{IE} . \tag{7b}$$

Equation (7a), vacancies in undersaturation, is highly probable in the stacking fault growth condition. However, eq. (7b) may not be accepted because: C_I is changed by oxygen solubility, since the redissolution of oxide precipitates occurs in the high temperature second annealing.[9] This means that C_I dereases inversely when the second annealing temperature increases. However, C_{IE} increases directly with temperature. This consideration suggests that bulk stacking fault growth is dominated by vacancies. Consequently, the right and left hand sides of eq. (4) can be written to be

$$CD/C_S = (1 - C_V/C_{VE})C_{VE}D_V/C_S = C_{VE}D_V/C_S , \tag{8a}$$

$$D_S = C_{VE}D_V/C_S . \tag{8b}$$

Equation (8b) indicates that the stacking fault shrinkage is also governed by vacancies.

From eqs. (3), (8a) and (8b), we obtain

$$C_{VE}D_V/C_S = 44 \exp(-4.4 \text{ eV}/kT) \text{ (cm}^2/\text{s)} . \tag{9}$$

In addition, difficulties arising from the self-interstitial model have been discussed in ref. 6. The difficulty is briefly reconsidered, as follows: According to the report, the diffusion coefficient of self-interstitials is expressed by

$$D_I = D_0 \exp(-4.4 \text{ eV}/kT) \text{ (cm}^2/\text{s)} . \tag{10a}$$

$$1.9 \times 10^6 \lesssim D_0 \lesssim 1.1 \times 10^7 . \tag{10b}$$

The upper limit of the diffusion coefficient ($D_0 = 1.1 \times 10^7$) is shown in Fig. 5. In addition, Seeger et al. estimated the diffusion coefficient of self-interstitials to be 1×10^{-6} cm^2/s at the melting point.[21] They considered denuded zone formation mechanism, i.e., outdiffusion kinetics of self-interstitials during cooling of FZ silicon growth. Temperature estimation is questionable in their estimation, but the data is utilized because of paper deficit. Mizuo and Higuchi have estimated it to be 1.1×10^{-9} cm^2/s at 1100°C.[22] They utilized lateral diffusion length of Boron which was diffused from an apperture

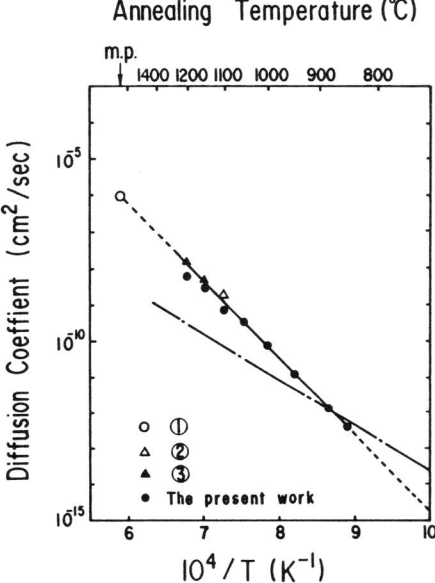

Fig. 5. Diffusion coefficients reported as that of self-interstitials. 1: ref. 21, 2: ref. 22, 3: ref. 23.

fabricated on silicon covered with Si_3N_4. Their estimation is based on the assumption that Boron mostly diffuses via self-interstitial diffusion, which might be accepted. Taniguchi et al. calculated it from stacking fault growth in surface layer induced by oxidation from rear side at 1100° to 1200°C.[23] These diffusion coefficients are shown in Fig. 5. It seems at first sight that the diffusion coefficient expessed by eq. (10a) agrees well with the previously reported values. However, the following contradiction arises: Oxide precipitate growth consists of both oxygen interstitial incoming diffusion, self-interstitial outgoing diffusion and vacancy incoming diffusion, as shown in Fig. 1(a). It is experimentally shown that the growth is limited by oxygen diffusion in the temperature range between 750° and 1050°C, which indicates that oxygen interstitial diffusion is the slowest process among them in the temperature range. However, the self-interstiti al diffusion coefficient is lower than oxygen interstitial diffusion coefficient when temperature is lower than 900°C, as shown in Fig. 5. Thus, this strongly suggests that the self-interstitial model is not valid. Here, there is a question why these data obtained by various methods are on a straight line in Fig. 5. As for the estimation of Seeger et al., the temperature esimation would be wrong, as described in this text. As for that of Mizuo and Higuchi, their data shows the lowest limit, and not the self-interstitial diffusion coefficient itself. Finally, the calculation by

Taniguchi et al. is based on the self-interstitial model and, therefore, it is worth to reconsidering their data by the vacancy model.

These independent two discussions on Figs. 4 and 5 lead a perspective in relationship between stacking fault growth and point defects that vacancies in undersaturation are the dominant driving force of stacking fault growth.

4. Bulk Stacking Fault Growth in the Transition Layer

4.1 Analysis of depth profile formation

The second assumption is made: *Point defect concentrations at the specimen surface reach their equilibrium concentrations at the beginning of the second annealing. They are then maintained throughout the annealing.* Considering the point defect concentration change under the specimen surface, eq. (2') can be modilfied to be

$$CD(z) = g_V(C_{VE}' - C_V(z))D_V + g_I(C_I(z) - C_{IE}')D_I . \qquad (12)$$

The in-diffusion profile of the vacancy concentration $C_V(z)$ and the out-diffusion profile of the self-interstitial concentration $C_I(z)$ can be given by[24]

$$C_V(z) = C_V + (C_{VE} - C_V)\,\text{erfc}(z/2(D_V t)^{1/2}) , \qquad (13a)$$

$$C_I(z) = C_I - (C_I - C_{IE})\,\text{erfc}(z/2(D_I t)^{1/2}) . \qquad (13b)$$

Here, the recombination effects of vacancy and self-interstitial on the depth profiles of these point defect concentrations are ignored. Substituting eqs. (13a) and (13b) into eq. (12),

$$CD(z) = g_V D_V (C_{VE} - C_V)\,\text{erf}(z/2(D_V t)^{1/2})$$

$$+ g_I D_I (C_I - C_{IE})\,\text{erf}(z/2(D_I t)^{1/2}) . \qquad (14)$$

The simplifications employed in eq. (5'), $C_{VE} = C_{VE}'$ and $C_{IE} = C_{IE}'$ are used again. Substituting eq. (14) into CD of eq. (1), and then integrating eq. (1), a final expression for the depth profile of the bulk stacking fault radius is obtained:

$$\int (1 - (r_c/r)^2)^{-1/2} (\ln(8r/r_c))dr$$

$$= 2\pi v_o b^{-1} g_V D_V (C_{VE} - C_V) \int \text{erf}(z/2(D_V t)^{1/2})dt$$

$$+ 2\pi v_o b^{-1} g_I D_I (C_I - C_{IE}) \int \text{erf}(z/2(D_I t)^{1/2})dt . \qquad (15)$$

Here, the result in the preceding section is now utilized: vacancies dominate the growth of the bulk stacking faults; $g_V=1$ and $g_I=0$. Thus, eq. (15) can be expressed as

$$\int (1 - (r_c/r)^2)^{-1/2}(\ln(8r/r_c))dr$$
$$= 2\pi v_0 b^{-1} D_V(C_{VE} - C_V) \int \mathrm{erf}(z/2(D_V t)^{1/2})dr , \qquad (16)$$

The unknown pre-integral term on the right-hand side of eq. (16) indicates the radius of the bulk stacking faults deep within the specimen bulk. Thus, the term can be eliminated by normalization by the radius of the bulk stacking faults deep within the bulk. Therefore, the depth profile becomes a simple function of the vacancy diffusion coefficient and the annealing time.

4.2 Experimental procedures

Specimens for obtaining the depth profiles of the bulk stacking fault radius were 2000 μm thick, 3-inch diameter, both side polished which were cut from an as-grown Czochralski silicon crystal, 5×10^{15} B cm^{-3} grown in our laboratory. The specimens were first-annealed at 850°C for 24 hrs. The introduced oxide precipitates were square-shaped, 1000 Å in half diagonal length and 10^{10} cm^{-3} in density. Then, the specimens were second-annealed at 1080°C for 3 h, at 1170°C for 50 min or at 1270°C for 30 min. Depth profile of the bulk stacking fault radius was measured by a successive etching method using the Wright solution: A given location of the specimen surface was observed after etching for a certain time and the diameter of the bulk stacking faults newly appearing was measured by optical microscope. The procedure was repeated to obtain the depth profiles of the bulk stacking fault diameter by considering the etching rate.

4.3 Results and discussion

Figure 6 shows the experimental results of the depth profiles of the bulk stacking fault radius.[10] The specimens were annealed at 1170°C for 50 min. It is found that the transition layer is clearly formed and that the layer width is independent of oxygen concentration. Here, calculations fit the experimental data very well. The fitting parameter is the diffusion coefficient of vacancies. The outdiffusion profiles formed at 1270° and 1080°C were almost the same as those in Fig. 6. The diffusion coefficients obtained by fitting are shown in Fig. 7. The temperature dependence of the diffusion coefficient is expressed to be

$$D_V = 257 \exp(-(2.84 \pm 0.66)\mathrm{eV}/kT) . \qquad (17)$$

The obtained diffusion coefficient is very different from those reported as the

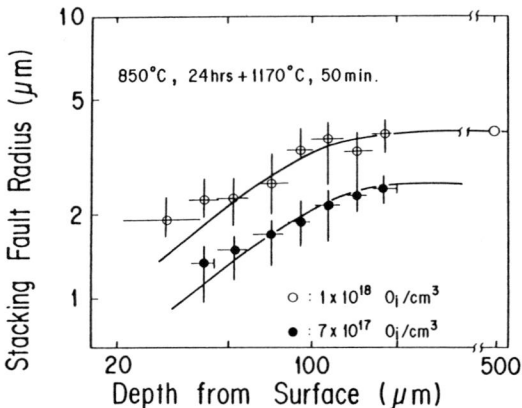

Fig. 6. Depth profiles of the bulk stacking faults under the specimen surface. The oxygen concentrations in these specimens are 7 and 10×10^{17} atoms/cm^3. The results are in good agreement with the calculation as shown by the solid lines.

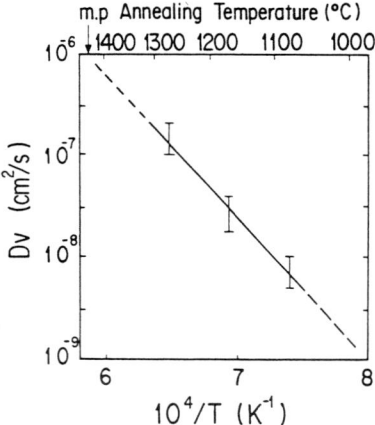

Fig. 7. Diffusion coefficient of the point defect responsible for the bulk stacking fault growth.

diffusion coefficient of self-interstitials shown in Fig. 5. This result together with the result obtained in the Section 2, strongly suggests that bulk stacking fault growth is drived by vacancies in undersaturation. Based on the cocept, the equilibrium concentration of vacancies can be calculated using eq. (9):

$$C_{\text{VE}} = 8.56 \times 10^{21} \exp(-(1.56 \pm 0.66)\text{eV}/kT) . \tag{18}$$

The concentration calculation is shown in Fig. 8.

This section successfully subdivides $C_{VE}D_V$ into the diffusion coefficient and equilibrium concentration. This is noteworthy since few papers on diffusion coefficient have so far reported. Therefore, comparisons are not readily available, which makes it difficult to evaluate the present results. One thing that should be remarked is that migration energy is fairly large 2.8 eV in comparison with that obtained by low temperature irradiation experiment. This might be the first evidence that vacancies would be extended at higher temperature suggested by van Vechten.[25] Further theoretical and experimental efforts should be made before thermal properties and nature of vacancies are fully understood.

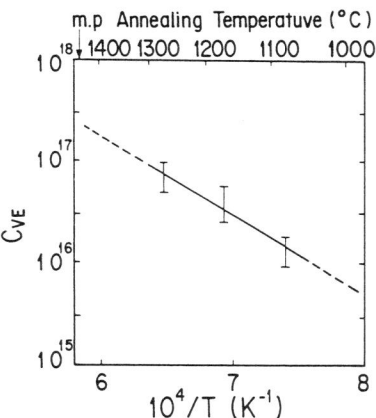

Fig. 8. Equilibrium concentration of the point defect responsible for the bulk stacking fault growth.

5. Vacancy and Self-Interstitial Components of Tracer Self-Diffusion Components

Tracer self-diffusion coefficient is expressed by

$$D_S = (f_V C_{VE} D_V + f_I C_{IE} D_I)/C_S . \qquad (19)$$

The uncorrelated vacancy component $C_{VE}D_V/C_S$ has been obtained in the present paper, eq. (9), as shown in Fig. 9. In addition, the uncorrelated self-interstitial component calculated from Au diffusion done by Morehead et al.[26] According to them, the self-interstitial component is shown by

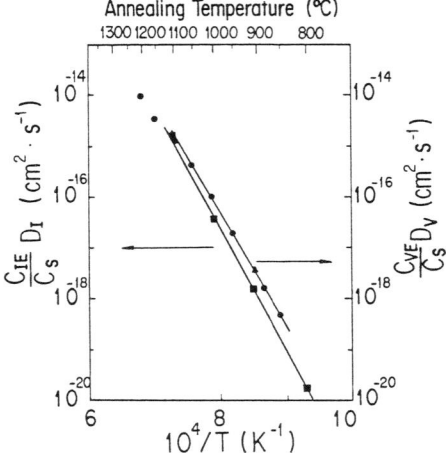

Fig. 9. Comparison between vacancy component of driving force of stacking fault growth and uncorrelated self-diffusion coefficient. $C_{VE}D_V = C_{IE}D_I$.

$$C_{IE}D_I/C_S = 918 \exp(-4.84 \text{ eV}/kT) . \qquad (20)$$

These components are also shown in Fig. 9. It is readily understood that

$$C_{VE}D_V \doteqdot C_{IE}D_I . \qquad (21)$$

It has been suggested that self-diffusion is governed by vacancies at temperature lower that around 1000°C,[27] which means that vacancy component has a low activation energy compared to self-interstitial component. This is in good acordance with the present result. Although similar result was obtained by Morehead et al., their result appears unreliable, since the vacancy component was estimated from Au diffusion by which they determined the self-interstitial component.[26] Therefore, the present result is the first result that the vacancy component of the self-diffusion coefficient is similar value to the self-interstitial component.

Physics behind the similarity is as yet unknown.

6. Conclusion

The growth kinetics of the stacking faults in specimen bulk and surface layer are studied, utilizing the two-step annealing technique. It has been found:

(1) The temperature dependence of the growth rate of bulk stacking faults is explained by that of self-diffusion coefficient.

(2) The dominant driving force of the bulk stacking fault growth is vacancies in undersaturation, not self-interstitials in supersaturation.

$$C_{VE}D_V/C_S = 44 \exp(-4.4 \text{ eV}/kT) .$$

(3) The diffusion coefficient D_V and equilibrium concentration C_{VE} are expressed by

$$D_V = 257 \exp(-(2.84 \pm 0.66)/kT) ,$$

$$C_{VE} = 8.56 \times 10^{21} \exp(-(1.56 \mp 0.66)/kT) .$$

(4) The vacancy component of uncorrelated self-diffusion coefficient is quite close to the self-interstitial component.

Acknowledgements

The authors would like to thank to Jiro Osaka for his valuable discussion.

REFERENCES

1) For example, C. P. Flynn: *Point Defects and Diffusion* (Clarendon Press, Oxford, 1972).
2) T. Y. Tan, E. E. Gardner, and W. K. Tice: Appl. Phys. Lett. **30** (1977) 175.
3) S. Mizuo: in this proceedings
4) S. M. Hu: J. Appl. Phys. **45** (1974) 1567; A. M. Lin, R. W. Dutton, D. A. Antoniadis, and W. A. Tiller: J. Electrochem. Soc. **128** (1981) 1121.
5) B. Leroy: J. Appl. Phys. **50** (1979) 7996.
6) For example, K. Wada, N. Inoue, and J. Osaka: *Defects in Semiconductors II*, ed. S. Mahajan and J. W. Corbett (Elsevier, New York, 1983) p. 107.
7) K. Wada, N. Inoue, and K. Kohra: J. Cryst. Growth **49** (1980) 749.
8) K. Wada, H. Takaoka, N. Inoue, and K. Kohra: Jpn. J. Appl. Phys. **18** (1979) 1629.
9) K. Wada and N. Inoue: J. Cryst. Growth **17** (1985) 111.
10) K. Wada and N. Inoue: to be published in J. Appl. Phys.
11) H. Takaoka, J. Osaka, and N. Inoue: Jpn. J. Appl. Phys. **18** (1979) Suppl. 18-1, 179; K. Wada and N. Inoue: *Defects and Radiation Effects in Semiconductors 1980*, Inst. Phys. Conf. Ser. 59 (1981) 461.
12) R. F. Peart: Phys. Stat. Sol. **15** (1966) K119.
13) R. N. Ghoshtagore: Phys. Rev. Lett. **16** (1966) 890.
14) J. M. Fairfield and B. J. Masters: J. Appl. Phys. **38** (1967) 3148.
15) H. J. Mayer, H. Mehrer, and K. Maier: *Lattice Defects in Semiconductors 1976* Inst. Phys. Conf. Ser. 31 (1977) 186.
16) L. Kalinowski and R. Seguin: Appl. Phys. Lett. **35** (1979) 211.
17) J. Hirvonen and A. Anttila: Appl. Phys. Lett. **35** (1979) 703.
18) F. J. Demond, S. Kalbitzer, H. Mannsperger, and H. Damjantschitsch: Phys. Lett. **93A** (1983) 503.
19) J. A. Lambert and P. S. Dobson: Philos. Mag. **A44** (1981) 1031.

20) U. Goesele and W. Frank: *Defects in Semiconductors*, ed. J. Narayan and T. Y. Tan (North-Holland, New York, 1981) p. 55.
21) A. Seeger, H. Foell, and W. Frank: *Radiation Effects in Semiconductors 1976*, Inst. Phys. Conf. Ser. 31 (1976) 12.
22) S. Mizuo and H. Higuchi: Jpn. J. Appl. Phys. **21** (1982) 272.
23) K. Taniguchi and D. A. Antoniadis: *Defects in Silicon*, ed. M. W. Bullis and L. C. Kimerling (The Electrochemical Soc, Pennington, 1983) p. 315.
24) H. S. Carslaw and J. C. Jaeger: *Conduction of Heat in Solids* (Oxford University Press, Oxford, 1959).
25) J. van Vechten: *Defects in Semiconductors*, ed. L. C. Kimerling and J. M. Parsey, Jr. (AIME, 1984).
26) F. Morehead, N. A. Stolwijk, W. Meyberg, and U. Goesele: Appl. Phys. Lett. **42** (1983) 690.
27) A. Seeger and K. P. Chik: Phys. Status Solidi **29** (1968) 276.

THE CHARACTERISTICS OF NITROGEN IN SILICON CRYSTALS

T. ABE, T. MASUI, H. HARADA, and J. CHIKAWA*

R & D Center, Shin-Etsu Handotai Co. Ltd., Isobe, Annaka, Gunma 379-01, Japan *National Laboratory for High Energy Physics, Tsukuba, Ibaraki 305, Japan*

Abstract. Nitrogen-doped crystals have been grown by the FZ and CZ methods. Their nitrogen concentrations in a range of 0.6 to 7×10^{15} atoms/cm^3 were measured by the infrared absorption and activation analysis. Observation of indentation rosettes showed that dislocation locking in N-doped crystals is stronger than that in conventional CZ crystals. Reduction of the infrared absorption peak due to nitrogen by annealing was much larger for the CZ crystals than for the FZ crystals. This implies that nitrogen atoms interact strongly with oxygen impurity and act as nucleation centers of oxygen precipitation. Nitrogen impurity was found to prevent generation of swirl and D-defects in FZ crystals. These observations may lead to understanding formation of microdefects.

1. Introduction

Dislocation generation in device processing often results in warpage of the wafers, which causes a difficulty in lithography. Wafers with large diameters suffer from stronger thermal stress during heat treatment such as oxidation and diffusion processes. Furthermore, the high degree of integration has been accompanied with the device structure that causes stress concentration, *e.g.*, multilayer metallization, U-shaped isolation, and trench structure. Therefore, mechanical strength of wafers should be increased, and CZ wafers with high oxygen concentrations have been used for integrated circuits. Recently, however, disadvantage of oxygen impurity has been widely recognized: To achieve a high degree of integration, SiO$_2$ films for the gate of MOS-FET must be thin, and its breakdown voltage will be the most important problem. The higher breakdown voltages have been obtained with decreasing oxygen concentration in crystals.[1] Oxygen microprecipitates are formed uniformly and may degrade quality of SiO$_2$ films. Furthermore, wafers with high oxygen concentrations are at the risk of warpage; if SiO$_2$

precipitation takes place, dislocations are generated easily from the precipitates. Therefore, strength of wafers should be obtained with oxygen concentrations as low as possible.

From this point of view, the present authors showed that nitrogen in silicon has a strong effect to lock dislocations using nitrogen-doped floatzoned crystals.[2,3] Recently, Chiou et al.[4] reported on the nitrogen effect for N-doped CZ crystals. Watanabe et al.[5] have grown oxygen-free CZ crystals using a crucible coated with Si_3N_4.

Nitrogen in silicon has interested some investigators. Kaiser et al.[6] doped nitrogen by reaction of N_2 or NH_3 gas with silicon at high temperatures. Yatsurugi et al.[7] found the solubility of nitrogen in the silicon melt (6×10^{18} atoms/cm^3), solid solubility (4.5×10^{15} atoms/cm^3), and equilibrium segregation coefficient (7×10^{-4}) by dissolving Si_3N_4 into the silicon melt. Chaney et al.[8] measured the dissolving rate of Si_3N_4 into the silicon melt. Nitrogen impurity was detected by photoluminescence (PL)[9] and deep level transient spectroscopy (DLTS).[10] It was found by EPR technique using N^+-implanted silicon specimens that nitrogen atoms are incorporated substitutionally in the silicon lattice.[11]

The purpose of this paper is to present some knowledge on nitrogen impurity which is required for practical use of N-doped material.

2. Experimental

2.1 Doping of nitrogen

Nitrogen-doped crystals were grown by the conventional FZ and CZ methods. In the case of the FZ method, the crystals were grown in an argon atmosphere added with nitrogen. Even if the atmosphere was 100% nitrogen, nitrogen did not react with the Si melt, but was very active with the Si solid at temperatures higher than 1300°C. Nitride films were formed on the polycrystalline silicon adjacent to the melt zone and dissolved into the melt by the zone travelling. The percentage of nitrogen in the atmosphere was a few % so that the nitrogen concentration in the silicon melt was less than the solubility limit at the melting point (1415°C). In the CZ method, therefore, Si_3N_4 was added in the silicon melt. Nitrogen in the melt hardly evaporated. However, when nitrogen concentration exceeded the solubility limit in the melt, Si_3N_4 was formed and floated on the melt surface. Its amount was so adjusted that the concentration in the crystal reached the solubility limit in the final stage of the growth.

2.2 Measurement of nitrogen concentrations

Ion-implanted nitrogen has been detected by secondary ion mass spectrometry (SIMS). However, the detection limit was found to be 3.2×10^{16} atoms cm^{-3} by using N^+-implanted specimens. This value exceeds the solubility limit in silicon crystals at the melting point. Therefore, this technique cannot be applied to the present experiment.

Infrared (IR) measurement was found to be a convenient method to detect nitrogen N as well as oxygen O and carbon C in silicon. Nitrogen is situated between C and O in the periodic table, and IR absorption peaks due to nitrogen was found at wave numbers of 963 and 764 cm^{-1}, as seen in Fig. 1. They are located between the oxygen peak at 1106 cm^{-1} and the carbon peak at 606 cm^{-1}. The peak at 963 cm^{-1} has a higher signal-to-noise ratio, and its peak height was used for concentration measurement: The IR absorption was measured for double-side polished wafers with a thickness of 2 mm. The calibration curve for the IR absorption in Fig. 2 was obtained by the Charge Particle Activation Analysis (CPAA) using ^{3}He particles which was developed by Yatsurugi et al.[7]; nitrogen concentration is given by the formula[12]

$$[N] = 1.82 \times 10^{17} \times \text{(absorption coefficient at 963 cm}^{-1}\text{)}.$$

The detection limit of this method was found to be 2×10^{14} atoms cm^{-3}.

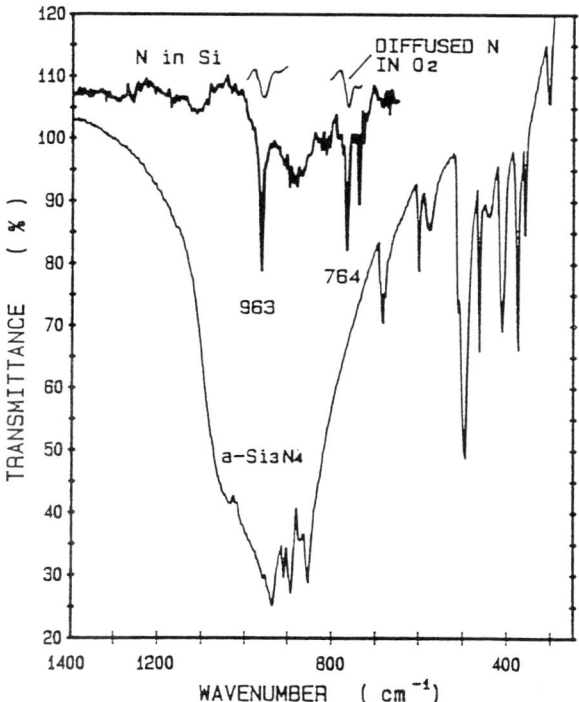

Fig. 1. Infrared absorption due to nitrogen in a FZ silicon crystal.

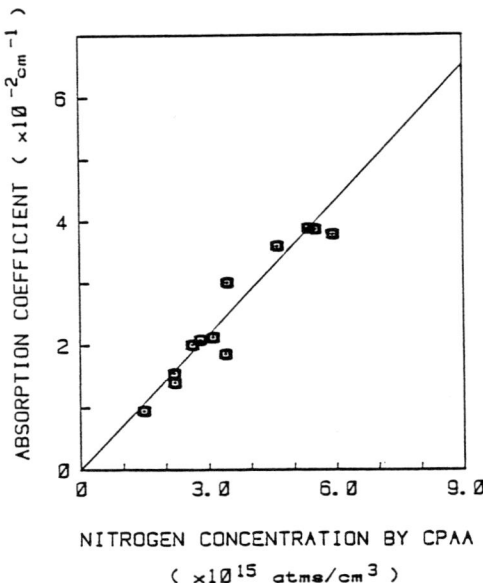

Fig. 2. Calibration curve for nitrogen concentration by the IR absorption measurement at the wave number of 963 cm^{-1}. See Fig. 1.

Photoluminescence (PL) and deep-level transition spectroscopy (DLTS) can be also used for measurement of nitrogen concentrations: Both the methods detect electrons bound around nitrogen atoms: DLTS reveals two donor levels due to nitrogen impurity, and luminescence due to nitrogen was also observed.

3. Nitrogen Effect on Dislocation Motion

Nitrogen in silicon was found to have effect to prevent slippage of silicon wafers. First, this effect will be compared with the oxygen effect. Next, the effect by coexistence of nitrogen and oxygen will be shown with N-doped CZ crystals.

3.1 N-doped FZ crystals

Concentration dependence of the nitrogen effect was examined by measuring sizes of rosette patterns due to indentation. The specimen crystals were indented for 30 s at 900°C by a Vickers indenter with a load of 0.98 N. The rosette patterns of dislocation etch pits around the indentations were revealed by etching with the Sirtl etchant. The result is shown in Fig. 3. It is seen that rosette expansion is smaller in N-doped FZ crystals than in CZ

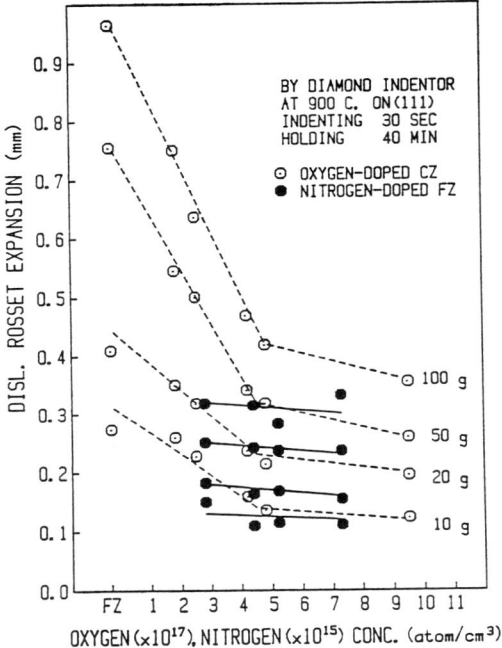

Fig. 3. Size of indentation rossettes vs oxygen or nitrogen concentration in silicon crystals.

crystals under the same indentation. The nitrogen effect is nearly constant in a wide range of nitrogen concentration (also see Fig. 4), whereas the oxygen effect expires in the concentration range less than 4.5×10^{17} atoms cm^{-3}. These observations indicate that N-doped FZ wafers are strong enough to prevent warpage by heat treatment in device fabrication.

3.2 Nitrogen effect on slippage

Three kinds of (111) wafers were prepared from a CZ crystal with an oxygen concentration of 1.7×10^{18} cm^{-3}, a conventional FZ crystal, and a nitrogen-doped FZ crystal with a concentration of 1.5×10^{15} cm^{-3}. The surfaces of these wafers were scratched in a $<110>$ direction with diamond stylus with a weight of 10 g. They were put into a furnace preheated at 1150°C by an automatic handler (room temperature to 1150°C for 5 min), kept for 10 min, and taken out to cool down to room temperature and then the scratched surface was etched by 10 μm to remove strains due to the scratch. X-ray topographs were taken as shown in the left of Fig. 4 and then, the same heat-treatment was made again and topographs were taken (the right of Fig. 4). During the heat treatment the stresses are applied radially, i.e., the stress is

EFFECTS OF NITROGEN ON DISLOCATION NUCLEATION AND MOVEMENT IN SILICON

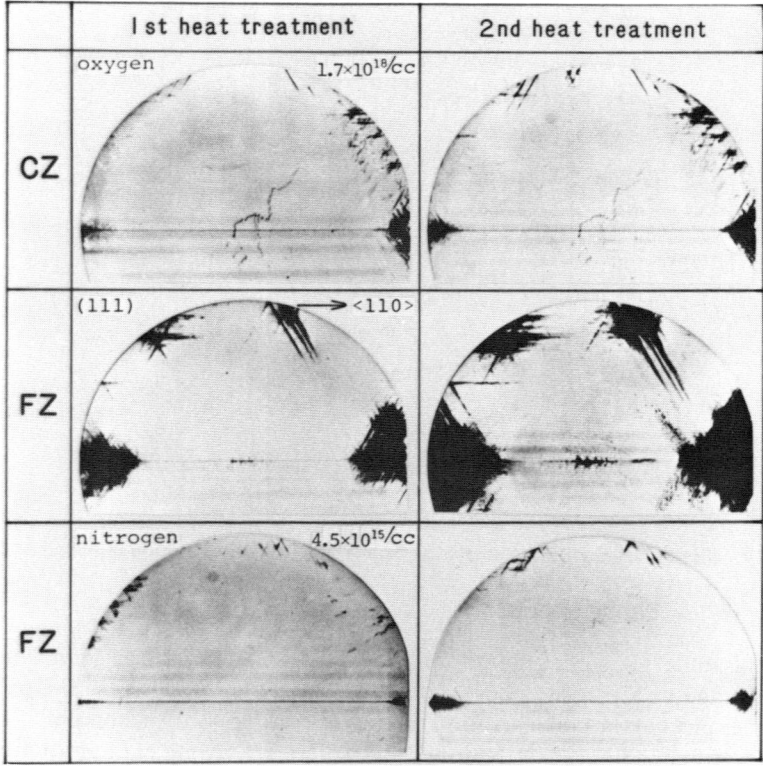

Scratched conditions

- Crystal diameter: 3 inches
- Diamond weight: 10g
- Scratching speed: 10 mm/sec

In diffusion funrace

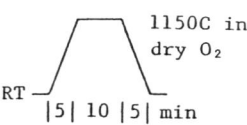

1150C in dry O_2

Fig. 4. X-ray topographs (220 ref. MoKα_1) obtained from three kinds of (111) wafers prepared from a CZ with an oxygen concentration of 1.7×10^{18} cm^{-3}, a conventional FZ crystal, and a nitrogen-doped FZ crystal with a concentration of 1.5×10^{15} cm^{-3} (compare with 4×10^{-2} cm^{-1} absorption coefficient). Left and right hand side topographs were taken after 1st and 2nd heat treatment, respectively, under the same condition.

higher at the periphery. The topographs in Fig. 4 show that dislocation generation is easier in an order of FZ≫CZ>N-doped FZ. The dislocation movement measured from comparison of the left and right topographs indicated that dislocation locking is stronger in an order of N-doped FZ~CZ≫FZ. These observation shows that N-doped FZ wafers are strong enough not to be warped by heat treatment in device fabrication.

3.3 N-doped CZ crystals

Figure 5 shows the dislocation Rosette expansion from indentations as function of the oxygen concentrations in the CZ and nitrogen-doped CZ (N-CZ) crystals. Under 10 ppma oxygen ($\times 1.6$=OLD ASTM), the toughness of CZ crystals against dislocation generation is decreased significantly, and over 10 ppma range, increasing with oxygen concentration increase. In case of N-CZ crystals, the toughness depends on not oxygen concentration but nitrogen. X-ray topographs of Fig. 6 shows the slippage distributions of 100 mm diameter FZ, CZ and N-CZ wafers after an epitaxial processing. It is clearly demonstrated that nitrogen atoms in silicon crystals even with two orders of maguitade lower concentrations compared to oxygen have strong locking effects for dislocation generation and movement.

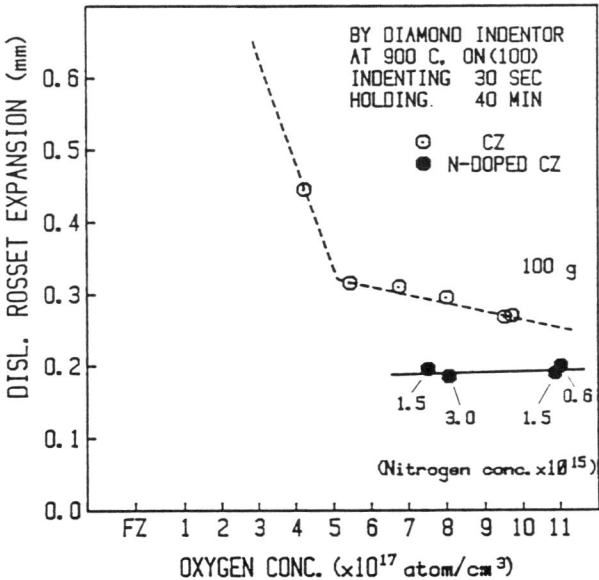

Fig. 5. Size of indentation rossettes vs oxygen concentrations measured for non-doped and nitrogen-doped CZ crystals. The nitrogen concentrations are indicated for each plot.

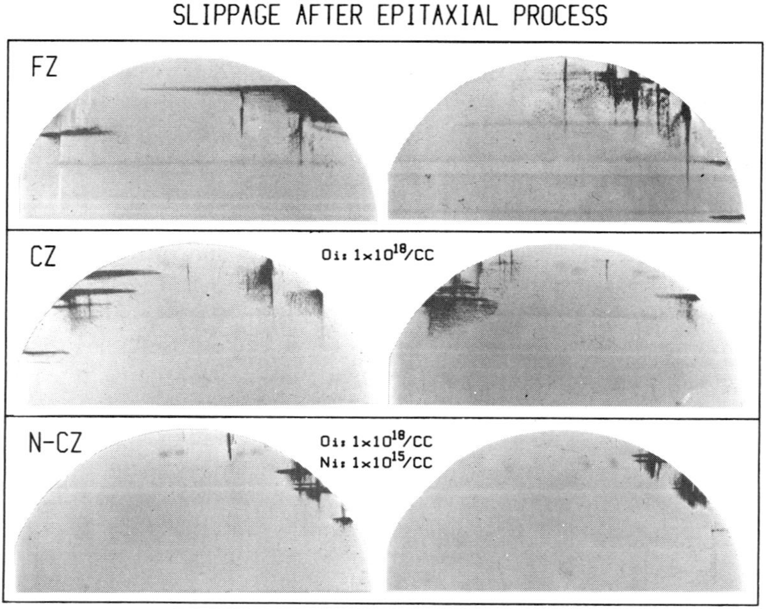

SLIPPAGE AFTER EPITAXIAL PROCESS

Fig. 6. X-ray topographs of FZ, CZ and nitrogen-doped CZ wafers which were processed with epitaxial growth. Slippages of each two wafers but just half parts are shown. All crystals are in <110> direction and 100 mm in diameter.

4. Nitrogen with Existence of Oxygen Impurity

To investigate interaction between nitrogen and oxygen in silicon, nitrogen concentration in a CZ crystal was measured by the IR and CPAA methods described in 2. The latter method detects the total amount of nitrogen impurity, while the IR absorption is due to individual nitrogen atoms accommodated in the silicon lattice. Figure 5 shows nitrogen concentrations measured along the growth direction by both the methods; the concentrations were plotted against the fraction frozen L. The dotted curve gives the concentration C_s calculated with the segregation coefficient $k=0.0007$[7] by

$$C_S = kC_0(1-L)^{k-1}, \qquad (1)$$

where C_0 is the initial nitrogen concentration in the melt. The concentrations measured by CPAA agree with the calculated values. Whereas, the concentrations obtained with the IR method are much lower. Since nitrogen concentra-

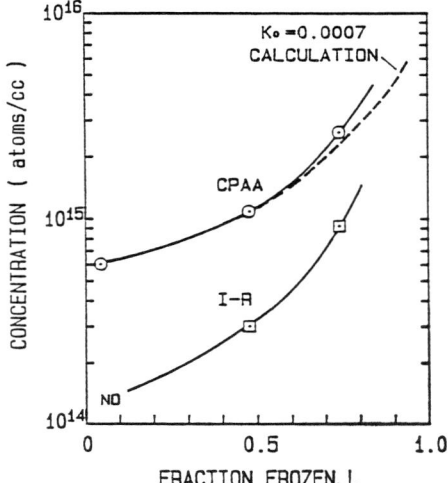

Fig. 7. Nitrogen concentrations measured along the growth direction of a nitrogen-doped CZ crystal. The dotted line was calculated by eq. (1) using the equilibrium segregation coefficient of $k_0 = 0.0007$.

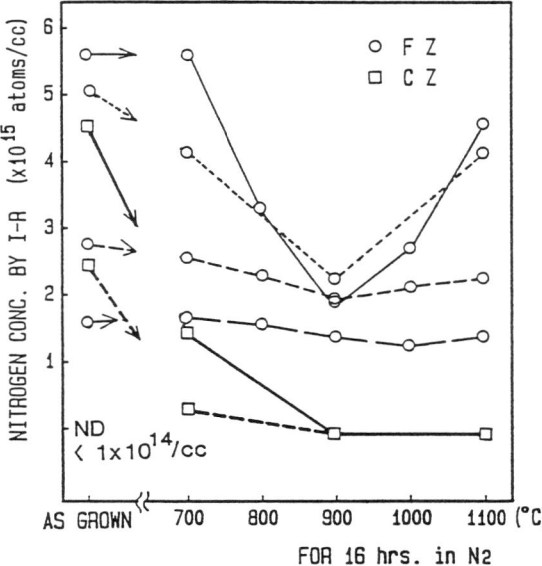

Fig. 8. Annealing effect on nitrogen concentrations detected by the IR absorption measurement (wave number 963 cm^{-1}). Annealing was made in a nitrogen ambience for 16 hrs at different temperatures.

tions in as-grown FZ crystals measured by both the methods agreed, this observation suggests that nitrogen atoms form some complexes and/or precipitates with oxygen atoms in CZ crystals.

Annealing effect on the IR absorption due to nitrogen was also investigated for FZ and CZ crystals. The specimens crystals were annealed at different temperatures for a fixed time of 16 hours, and then the IR absorption due to nitrogen was measured as shown in Fig. 6. The absorption or nitrogen concentration obtained by the IR method decreases by annealing. For FZ crystals, the decrease is maximum at 900°C, when nitrogen concentration is higher than 2×10^{15} atoms cm^{-3}. This observation may be explained as due to nitrogen precipitation which is controlled by supersaturation and diffusion of nitrogen atoms at the annealing temperature. For CZ crystals, however, the IR absorption disappears by annealing at around 900°C and the higher temperatures. This observation suggests that nitrogen atoms act as a nucleation centers of oxygen clusters and/or precipitates.

5. Nitrogen Effect on Formation of Microdefects

It is well known that two kinds of microdefects, swirls[13] and D-defects,[14-16] exist in the conventional FZ crystals. Swirls and D-defects were considered to be formed by supersaturated self-interstitials and vacancies, respectively. Recently, the present authors[15] found a new type of very small defects in regions free from both swirl and D-defects. They were referred to as "I-defects" and were attributed to microprecipitates of residual oxygen impurity in FZ crystals. Therefore, effect of nitrogen on the microdefect formation was examined by doping nitrogen during FZ growth.

FZ crystals with a diameter of 42 mm were grown with changing the growth rate. Specimens with a thickness of about 1 mm were cut longitudinally from these crystals. After copper-decoration, they were observed by X-ray topography, as shown in Fig. 9. In Fig. 9(a), the growth rate was

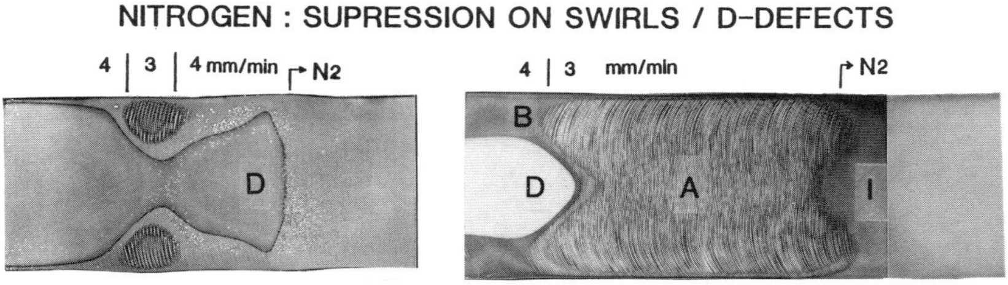

NITROGEN : SUPRESSION ON SWIRLS / D-DEFECTS

Fig. 9. Nitrogen effect on formation of micro defects in FZ crystals. Crystal diameter: 42 mm. The growth rate was changed and doped with nitrogen during the growth as indicated. (a) nitrogen effect on D-defects (vacancy clusters), (b) nitrogen effect on swirl defects.

changed from 4 to 3 and 3 to 4 mm/min, and then nitrogen was doped. The region labelled "D" is seen from rim to rim in the initial part grown at 4 mm/min and becomes narrower in the region grown at 3 mm/min, where swirl defects labelled "S" are seen in the periphery. The region "D" becomes thicker again by increasing the growth rate to 4 mm/min, but disappears by doping nitrogen. As described in the previous paper,[15] D-defects are distributed uniformly in D-regions, and swirl defects (S-regions) are formed in the regions grown at 3 mm/min. I-defects exist everywhere in the rest. The D-region is bordered with a black line or band which has a low X-ray diffracted intensity, i.e., the edge of a D-region is perfect, free from D-defects and I-defects. As seen from Fig. 9(a), the D-region suddenly disappears by doping nitrogen, and I-defects are formed. In Fig. 9(b), the growth rate was changed from 4 to 3 mm/min. D-defects and swirl defects in striated distribution are seen before and after the change, respectively. By doping nitrogen, it is seen that swirl defects change into an I-region (I-defects). In Fig. 9(c), the initial region is similar to the case of Fig. 9(a). By doping oxygen, the I-regions in the periphery become wider, and D-defects are seen in striated distribution. By doping nitrogen in addition, the D-region changed into the I-region. These observations show that doping nitrogen always changes the formation patterns of microdefects into that of I-defects, i.e., nitrogen impurity strongly assist to form I-defects. This phenomenon was found to take place even at such low concentrations of nitrogen that are undetectable by the IR absorption measurement.

The disappearance of both swirl and D-defects by doping nitrogen at extremely low concentrations cannot be explained by the theories that assume supersaturation of self-interstitials or vacancies.[17,18] In the previous paper,[15] we interpreted formation of microdefects as follows: Vacancies play a principal role on their formation and interactions between vacancies and impurities are very important; oxygen solubilities are enhanced by increasing concentration of supersaturated vacancies. Consequently, vacancy-controlled precipitation of residual oxygen impurity is responsible for formation of A-, B-, and I-defects: one of these types of defects is formed by oxygen impurity remaining in FZ crystals, depending upon the temperature that out-diffusion of vacancies depending on both the growth rate and temperature gradient at the interface takes place at. Swirl defects (A and B defects) are formed by oxygen precipitation at the nucleation sites formed by remelting during the growth. This takes place at relatively high temperatures where oxygen can diffuse. I-defects are formed in relatively low temperatures where supersaturation of oxygen impurity is high. In the light of the experimental result described in 4, one can conclude that nitrogen atoms interact strongly with oxygen impurity; this interaction prevents the oxygen diffusion and acts as nucleation centers for oxygen microprecipitates (I-defects).

6. Conclusion

High strength of wafers was achieved by doping nitrogen at a concentration level as low as 10^{15} atoms cm^{-3}. It is likely that dislocation locking is due to complexes of nitrogen and oxygen atoms (In FZ crystals, residual oxygen of 10^{16} atoms cm^{-3}). Nitrogen at such low concentrations is expected to have minor effect on electrical properties. Nitrogen-doped crystals are proposed as new material for fabrication of VLSI.

REFERENCES

1) F. Kiyosumi, H. Abe, H. Sato, M. Ino, and K. Uchino: Trans. Inst. Electro. Comm. Eng. Jpn. **SSD** 83-86 (in Japanese).
2) T. Abe, K. Kikuchi, S. Shirai, and S. Muraoka: *Semiconductor Silicon 1981*, ed. H. R. Huff *et al.* (Electrochem. Soc., Pennington, 1981) p. 54.
3) K. Sumino, I. Yonenaga, M. Imai, and T. Abe: J. Appl. Phys. **54** (1983) 5016.
4) H. D. Chiou, J. Moody, R. Sandfort, and F. Shimura: Electrochem. Soc. 1984 Spring meeting, Extended Abstract No. 36.
5) M. Watanabe, T. Usami, H. Muraoka, S. Matsuo, Y. Imanishi, and H. Nagashima: *Semiconductor Silicon 1981*, ed. H. R. Huff *et al.*(Electrochem. Soc., Pennington, 1981) p. 126.
6) W. Kaiser and C. D. Thurmond: J. Appl. Phys. **30** (1959) 427.
7) Y. Yatsurugi, N. Akiyama, Y. Endo, and T. Nozaki: J. Electrochem. Soc. **120** (1973) 975.
8) R. E. Chaney and C. V. Varker: J. Electrochem. Soc. **123** (1976) 846.
9) M. Tajima, T. Masui, T. Abe, and T. Nozaki: Jpn. J. Appl. Phys. **20** (1981) L423.
10) Y. Tokumaru, H. Okushi, T. Masui, and T. Abe: Jpn. J. Appl. Phys. **21** (1982) L443.
11) T. B. Mitchell, P. P. Pronko, J. Shewchun, D. A. Thompson, and J. A. Davis: J. Appl. Phys. **46** (1975) 332.
12) Y. Itoh *et al.*: to be published.
13) A. J. R. de Kock: Philips Tech. Repts. Sppl. No. 1, (1973) 1.
14) K. J. Roksnoer and M. M. B. Van den Boom: J. Cryst. Growth **53** (1981) 563.
15) T. Abe, H. Harada, and J. Chikawa: *Defects in Semiconductors II*, ed. S. Mahajan *et al.* (North-Holland, NY., 1983) Vol. 14, p. 1.
16) W. Keller and A. Mühlbauer: Inst. Phys. Cong. Ser. **23** (1975) 538.
17) V. V. Voronkov: J. Cryst. Growth **59** (1982) 625.
18) V. Gösele and T. Y. Tan: *Aggregation Phenomena of Point Defects in Silicon*, ed. E. Sirtl *et al.* (Electrochem. Soc., Pennington, 1982).

OXYGEN IN SILICON

N. INOUE, K. WADA, and J. OSAKA

NTT Atsugi Electrical Communications Laboratories, 3-1 Morinosato Wakamiya, Atsugi, Kanagawa, 243-01 Japan

Abstract. Oxygen is present in most silicon used for device fabrication and plays an important role in both material and device properties. Consequently, the control of oxygen and oxygen precipitation is becoming more important for state-of-the-art integrated circuits. This paper reviews the current status of several important aspects of oxygen in silicon. Applications of this oxygen, which here we call "defect engineering", are also shown. Furthermore, the directions for future research will be discussed.

1. Introduction

Currently, the majority of the silicon crystal used for electronic device fabrication is grown by the Czochralski (CZ) method. Because the crystals are pulled from silica crucibles, they contain a large amount of dissolved oxygen (about 1×10^{18} atoms/cm^3). Unfortunately, this oxygen in silicon has been responsible for many harmful effects. However, it has been recently recognized that oxygen can also have beneficical effects, such as wafer hardening and intrinsic gettering. By controlling the oxygen content and its behavior, these beneficial effects and reduction in detrimental effects may be achieved. Therefore, much effort has been devoted to the understanding of oxygen kinetics in silicon.[1-3] This paper reviews the present status of several important aspects of oxygen in silicon, such as oxygen incorporation, precipitation, and electronic behavior. Their applications, which here called "defect engineering", are also shown. Furthermore, the directions for future research will be discussed.

2. Oxygen Incorporation

Oxygen incorporation is a three-step process; quartz crucible dissolution, transportation in silicon melt accompanied by evaporation, and solidification. The crucible dissolution rate is about 1×10^{-6} g/cm^2·s under a representative growth condition and depends mainly on the melt temperature and fluid flow velocity.[4] Oxygen transportation and evaporation determine the nonuniform melt oxygen concentration. It has been found that as much as 99% of the dissolved oxygen escapes from the melt surface in the form of SiO vapor.[5] The most striking effect of convection on oxygen concentration is demonstrated by a remarkably low oxygen content crystal obtained by growth under a magnetic field which supresses convective fluid flow.[6] It is well known that the melt aspect ratio (fraction of melt solidified), crystal/crucible diameter ratio, and ambient pressure also affect the oxygen concentration.

It has not yet been established if the oxygen segregation coefficient at the solid liquid interface is greater or less than unity. Thus far, an equilibrium coefficient of 1.25 has been reported and widely accepted.[7] However, studies on the oxygen striated distribution due to a periodic growth rate change suggested that this coefficient is about 0.25, which is much smaller than unity.[8] Because of oxygen evaporation, it is difficult to accurately determine the segregation coefficient.

Precise oxygen control, both in concentration and distribution, is now necessary for "defect engineering". The magnetic field-applied Czochralski method yields crystals with an oxygen content as low as 1×10^{17} atoms/cm^3. These crystals are successfully used for CCD devices which require zero defects in the device area.[6] More widely used is to optimize crystal and crucible rotations.[9]

3. Fundamental Properties of Oxygen in Silicon

Oxygen in silicon has an interstitial configuration. Uncertainties still exist in the fundamental properties of oxygen in silicon, such as solubility and diffusion coefficient. Infrared absorption band at 9 μm (1107 cm^{-1}) has been identified with the antisymmetric vibrational mode of the bent Si–O–Si unit.[10] The infrared absorption band has been correlated with the oxygen content by various techniques. The calibration factor for the room temperature peak absorption coefficient varies about a factor of two between authors. Recent determination by JEIDA (Japan Electronic Industry Development Association)[11] by charged particle activation analysis is given by;

$$[O] = 3.03 \times 10^{17} \text{ atoms} \cdot \text{cm}^{-3}/\text{cm}^{-1}. \tag{1}$$

This value is supported by more recent research.

The solubility of oxygen in silicon reported by Hrostowski and Kaiser[12] has been widely used. It is 1.8×10^{18} atoms/cm^3 near the melting point and the enthalpy of solution is 0.95 eV (22 kcal/mole). Recent work of Craven[13] gives a solubility curve nearly identical to the above report (corrected for his calibration constant of 4.815×10^{17} atoms·cm^{-3}/cm^{-1} to that given in eq. (1)),

$$C_E = 2.00 \times 10^{21} \exp(-1.032 \text{ eV}/kT) \text{ atoms}/\text{cm}^3. \qquad (2)$$

This curve is shown in Fig. 1.

The diffusivity of oxygen in silicon can be determined by various techniques. The most widely used until recently are those of Takano and Maki[14] determined by lattice strain measurement between 1200 and 1100°C. Mikkelsen[15] obtained the diffusivity by SIMS in the wide temperature range of 700–1240°C, which supports the above data. Recently, Stavola et al.[16] reported anomalously fast oxygen atomic hopping at low temperatures in crystals not receiving the 1350°C heat treatment (dispersed). The diffusion coefficient determined from the dispersed sample given by the following equation agrees with earlier determinations, as shown in Fig. 2,

$$D = 0.091 \exp(-2.4 \text{ eV}/kT) \text{ cm}^2/\text{s}. \qquad (3)$$

4. Oxygen Precipitation

4.1 Structure of precipitates

There have been several structural studies of oxide precipitates from the viewpoints of morphology and crystallography with TEM. Depending on the

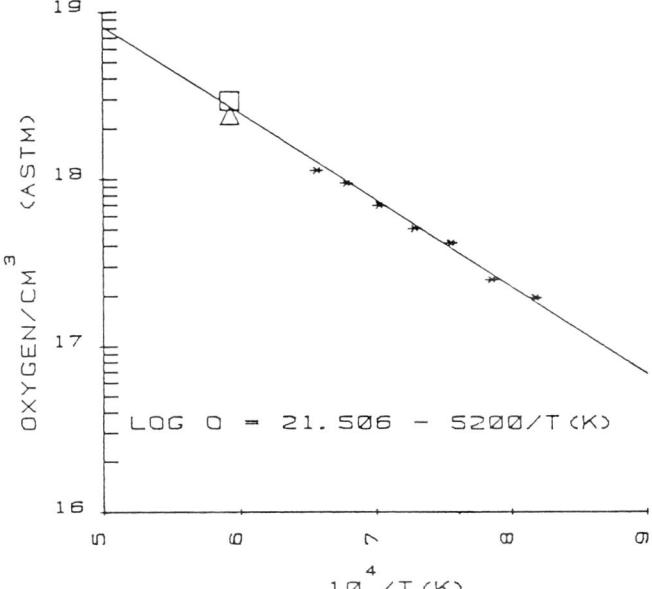

Fig. 1. Solubility of oxygen in silicon (ref. 13). The conversion factor is 4.815×10^{17} atoms·cm^{-3}/cm^{-1} (infrared absorption coefficient).

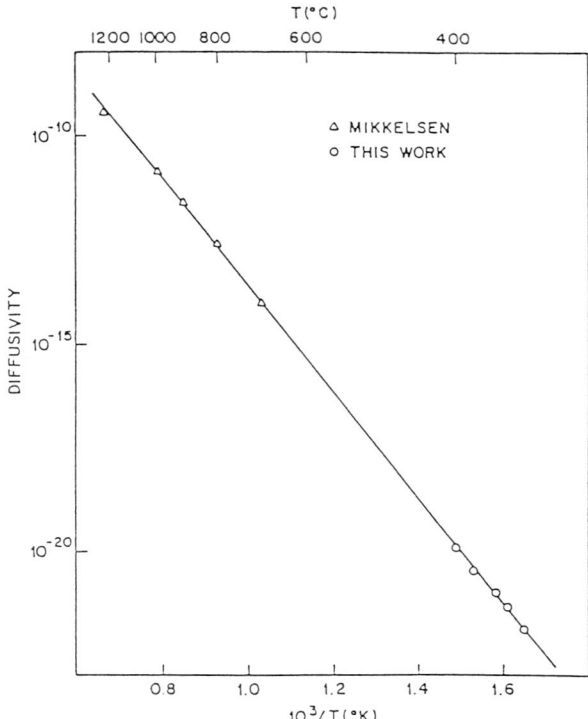

Fig. 2. Diffusivity of oxygen in silicon (ref. 16).

annealing temperature, three main features are obtained: First, near 800°C, square-shaped precipitates on the (100) planes with <110> edges are predominant as shown in Fig. 3.[17] Second, at temperatures above 1000°C octahedral precipitates bound by (111) planes are also observed.[18] Finally, at 650°C, rod-like defects on (100) planes along the <110> directions are formed.[19] Microdiffraction as well as structural images clarified that the former two features are amorphous whereas the last is coesite crystal, which is a dense high pressure form of silica.[19]

4.2 Factors affecting oxygen precipitation

At first, oxygen precipitation in silicon seems a very simple phenomenon. However, the number of factors affecting formation of oxide precipitates, either nucleation or growth, have increased as research in this area developed, listed as follows.

1) oxygen concentration, 2) annealing temperature and time, 3) carbon concentration, 4) thermal behavior of specimens, 5) annealing ambient, 6)

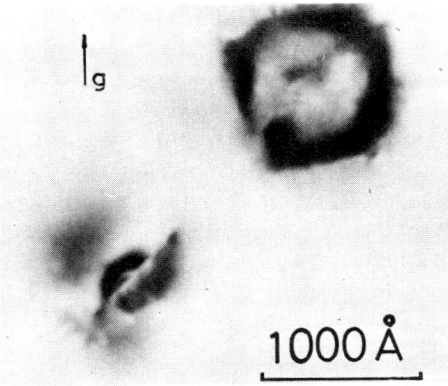

Fig, 3. Transmission electron micrograph of (100) platelet oxide precipitates (ref. 17).

dopant species and content, 7) wafer damage, 8) growth fluctuation. Not all of these factors are now completely established as described below.

4.3 Nucleation mechanism

Many investigations have been performed in order to understand the details of the oxide precipitate nucleation. There are two nucleation mechanisms, homogeneous and heterogeneous nucleation. It is particularly important to determine whether oxygen precipitation takes place via homogeneous nucleation or heterogeneous nucleation which involves other chemical impurities or structural imperfections as nucleation sites. In the following, two nucleation stages are separately discussed; nucleation during crystal growth and nucleation during annealing of grown crystals. This is because there is a significant difference in the thermal condition between these two stages.

(1)   Homogeneous nucleation during annealing

For nucleation during annealing, Freeland et al.[20] first proposed a homogeneous nucleation model. They measured the defect density after annealing at 1200°C in specimens with and without preannealing at 700°C. It was found that no defects were included in the latter whereas a large number of defects were formed in the former. They considered that these defects were nucleated at 1200°C, and that the effect of low temperature preannealing on decreasing the required supersaturation for nucleation suggested that homogeneous nucleation existed. The required supercooling was 80°C. Osaka et al.[21] have attempted to measure the nucleation rate directly. Their experiments utilized a two stage anneal in which specimens were first annealed for various periods of time at low temperature, and then subjected to a high temperature anneal. The latter caused the nucleated microprecipitates to grow to an observable size. A linear precipitate density increase with time was observed at 750°C as shown in Fig. 4, and the nucleation rate was found to be about $10^7/\text{cm}^3\cdot\text{s}$. Inoue et al.[2] obtained the dependences of nucleation rate on oxygen content and temperature as shown in Fig. 5. Little nucleation is observed above 950°C, contrary to speculation by Freeland et al. The required superstaturation for nucleation was determined to be about 300°C. It was likely that Freeland et al. observed the precipitates nucleated either during cooling after growth or during annealing at 700°C.

The following discussion shows that our results are well explained by the classical nucleation theory, which supports the homogeneous nucleation model. In the classical nucleation theory,[22] it is assumed that an equilibrium distribution of cluster sizes exists. The equilibrium "embryo" density is given by the following equations,

$$N(r) = N \exp(-G(r)/kT), \tag{4}$$

$$G(r) = (4/3)\pi r^3 \Delta G_v + 4\pi r^2 \sigma, \tag{5}$$

$$\Delta G_v = (\Delta F_v/T)\Delta T. \tag{6}$$

Here N is the oxygen concentration (assuming homogeneous nucleation at oxygen atoms), $\Delta G(r)$ is the free energy of formation of an embryo with radius r, ΔG_v is the molar free energy difference between oxide precipitate and solution, σ is the interfacial energy per unit area between precipitates and

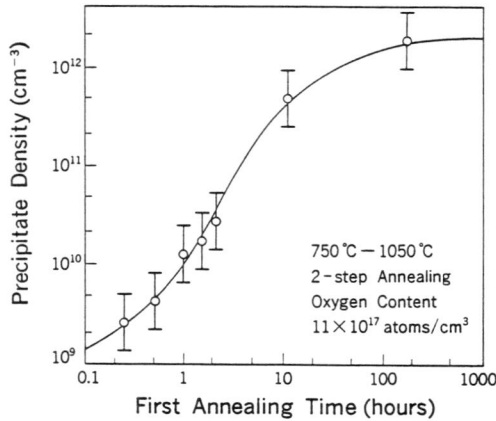

Fig. 4. Oxide precipitate density vs annealing time in silicon 2-step annealed from 750°C to 1050°C (ref. 21).

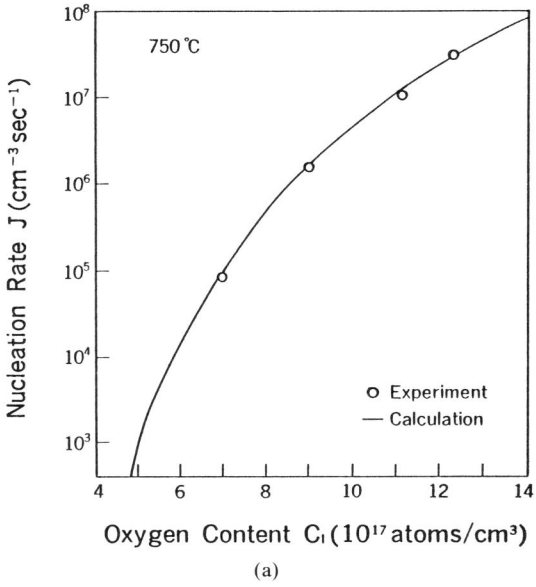

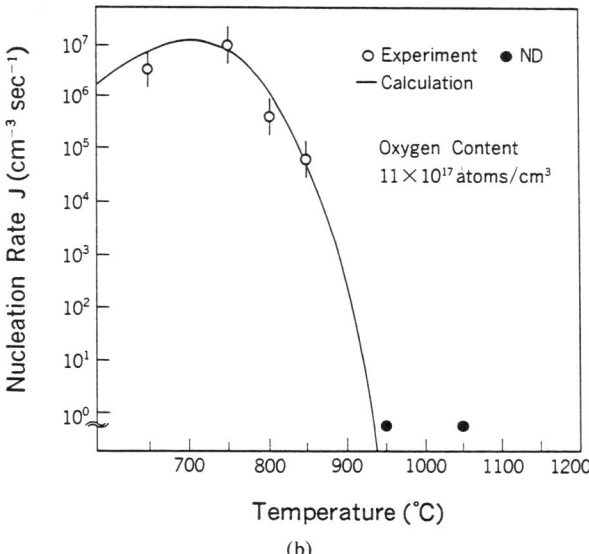

Fig. 5. (a) Oxygen content dependence of precipitate nucleation rate at 750° C and 850° C (ref. 2). (b) Temperature dependence of precipitate nucleation rate. Oxygen content is 1.2×10^{18} atoms/cm^3 (ref. 2).

matrix, ΔF_v is the dissolution enthalpy, and ΔT is the supercooling. Then, the steady state nucleation rate is given by,

$$J = Z \cdot N_c \cdot W. \tag{7}$$

In this equation N_c is the equilibrium critical nucleus density ($N(r_c)$ where r_c is critical radius), Z is the so-called Zeldovich factor expressing the ratio of steady state critical nucleus density to the equilibrium density, and W is the sticking rate of oxygen atom to the critical nucleus. The observed nucleation rate fitted well with eq. (7) using σ and Z as fitting parameters. This strongly suggests that homogeneous nucleation takes place below 950°C. The calculated equilibrium embryo size distribution is shown in Fig. 6. The steady state embryo size distribution is considered to be formed during continuous cooling after growth. Several other papers also suggest the homogeneous nucleation.[23]

(2) *Heterogeneous nucleation in annealing*

Many reports have suggested the existence of the heterogeneous nucleation mechanism. Most of them, however, have dealt only with nucleation during cooling after growth,[3,24] and few have dealt with nucleation during annealing.[25] In these reports, carbon atoms are assigned as heterogeneous nucleation sites, and the role of carbon will be discussed in Section 4.4(2).

(3) *Nucleation during cooling after crystal growth*

The results of many reports concerning this nucleation are summarized in

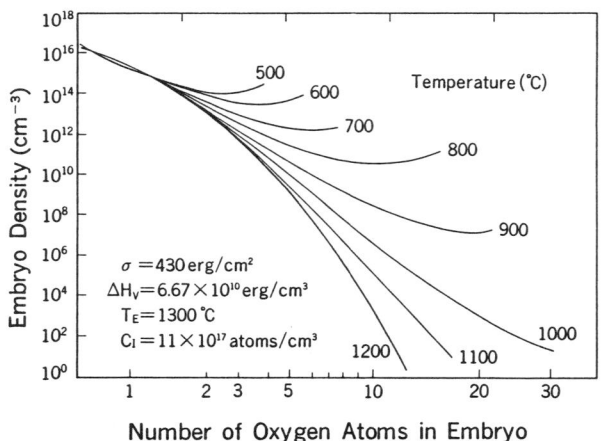

Fig. 6. Calculated equilibrium embryo density. The parameters are $\sigma=430$ erg/cm^2 and $Z=0.01$ (ref. 2).

Table 1. Ravi[26] proposed a heterogeneous nucleation model where vacancy clusters were postulated as nucleation sites. Next De Kock et al.[27] suggested that heterogeneous nucleation takes place predominantly near 1000°C during cooling after crystal growth, where silicon interstials play an important role, whereas homogeneous nucleation takes place below 800°C. We have found that there are two types of precipitates included in as-grown crystals. Microprecipitates (about 10Å in size) are determined from their size distribution to be nucleated homogeneously during cooling below 950°C.[28] The nucleation temperature of large precipitates (about 1000Å in size) has been determined to be near 1200°C and they have been determined to be hence formed through the heterogeneous nucleation mechanism.[29] This is because the homogeneous nucleation is negligible at such high temperatures.

During continuous cooling after growth, the metastable embryo (critical nucleus) at a particular time and temperature is converted into stable precipitates at a slightly lower temperature without absorbing an atom. In contrast to this situation, it is clear that the critical nucei can only become stable by absorbing an additional atom (nucleation) during annealing. Whether the above "spontaneous stabilization" or nucleation is dominant during cooling is not yet clear.

As seen from the above discussion, the nucleation mechanism is not unique, but changes with temperature. It is now widely accepted that the heterogeneous nucleation of oxide precipitates dominates at high temperature, whereas homogeneous nucleation dominates at low temperature.

4.4 Effects of other factors
(1) Oxygen thermal donors

It is well known that so-called "thermal donors" are induced into CZ silicon by annealing at temperatures near 450°C.[30] As described in Section 5,

Table 1. Models for oxide precipitate nucleation during cooling after crystal growth.

Temperature	Mechanism	Author
1200°C	Heterogeneous (Growth Fluctuation)	Wada[29]
?	Hetero. (Vacancy)	Ravi[26]
< 1000°C	Hetero. (Self Interstitials)	de Kock[27]
< 950°C	Homo.	Inoue[28]
< 800°C	Homo.	de Kock[27]
~ 650°C	Hetero. (Carbon)	Kishino[3]

their density is about $10^{16}/cm^3$, which is very high, and their configuration is accepted as SiO_x clusters. From infrared analysis, it has recently been established that the thermal donors are series of SiO_x clusters including 3 to 5 oxygen atoms. Furthermore, these clusters including more oxygen atoms are formed later.[31] Thermal donors disappear during annealing at near 650°C, which is thought to be due to the absorption of an additional oxygen atom, or change into precipitates including more than 5 oxygen atoms.[32] The above features are the same as those expected for embryos in homogeneous nucleation. In other words, it is very likely that the embryo size distribution necessary for homogeneous nucleation is formed by annealing, and that a part of it is detected as thermal donors through its electrical activity. The thermal donors are included in as-grown silicon, which supports this concept. Moreover, it has been reported that preannealing at 450°C enhances precipitate nucleation in the subsequent annealing at 750°C.[33] The presence of thermal donors represents strong evidence that homogeneous nucleation is possible in silicon.

(2) Carbon

Oxygen and carbon are the two main impurities in silicon crystals grown by the CZ method. Carbon in silicon is supplied from graphite hot-zone components in the crystal-growing furnace. The concentration of carbon in silicon ranges from less than 1 to 10×10^{16} atoms/cm^3. The role of carbon in relation to oxygen precipitation is highly controversial. Some experimentation suggests that a high carbon concentration can be very helpful in nucleation[3] as shown in Fig. 7. However, other results suggest that the correlation between carbon impurities and oxygen precipitation is very weak.[34] The heterogeneous nucleation model at carbon atoms are based on the following considerations.

1) The dissolved oxygen content reduces faster during annealing in carbon rich samples than in carbon lean samples.[3]

2) High temperature preannealing inhibits precipitation during subsequent low temperature annealing, which supports the viewpoint that homogeneous nucleation is impossible.[3]

3) Oxygen precipitation accompanies a reduction in substitutional carbon. The carbon diffusion coefficient is so small that carbon reduction is only possible when nucleation of precipitates occurs at carbon atoms.[35] However, there are several problems which exist in the above discussion such as:

1) With an indirect approach of measuring precipitated oxygen, it is not clear if enhanced nucleation or enhanced growth occurred. Moreover, precipitation occurs without nucleation, when precipitate decoration takes place on secondary dislocation loops formed at primary precipitates.

2) The correlation of precipitated oxygen content to carbon content is not sufficient. Slices from the same crystals with different carbon concentrations

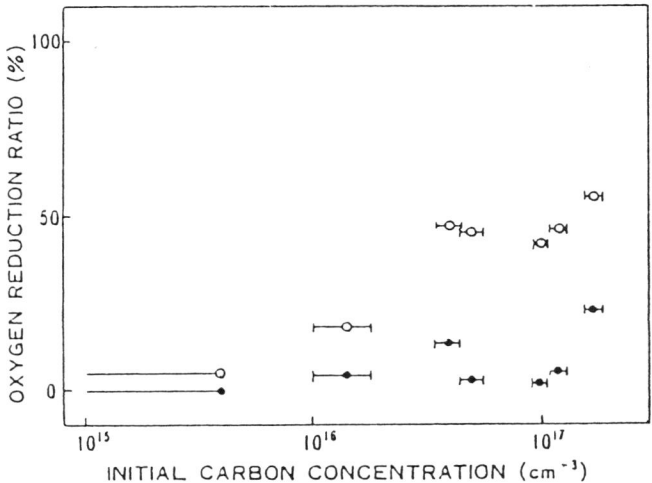

Fig. 7. Carbon concentration dependence of oxygen precipitation (ref. 3).

might differ in other respects. They include thermal behavior during growth or growth rate fluctuations which introduce possible nucleation sites other than carbon atoms.

3) Retarded but distinct oxide precipitate nucleation in high temperature preannealed samples has been found. This phenomenon demonstrates the "presence" of homogeneous nucleation rather than its absence as discussed in (3).

4) A substitutional carbon concentration reduction is possible without nucleation at carbon atoms, when supersaturated silicon interstitials are formed and replace substitutional carbon atoms. Actually, oxygen precipitation accompanies the emission of interstitials, as described in Section 4.6. Another possible explanation is that carbon diffusion may be enhanced by supersaturated silicon interstitials.

(3)   *Thermal history of specimens: time-lag in nucleation*

Oxygen precipitation kinetics change from ingot to ingot and from furnace to furnace. Kishino et al.[36] suggested that this is partially due to different thermal history of crystals received during cooling after crystal growth. This is expected considering the nucleation kinetics dependence on temperature as shown in Section 4.3. Precipitation in annealing is also affected by the preceding annealing procedure. Generally, low temperature preannealing enhances precipitation in the following high temperature annealing, as shown in Fig. 8.[37]

Precipitation retardation by high temperature preannealing has been pointed out by Kishino et al.[3] as shown in Fig. 9. Similar results were obtained

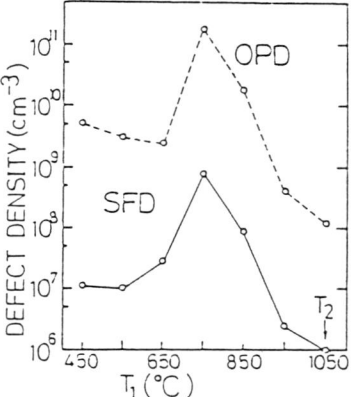

Fig. 8. Defect density after annealing at 1050°C increased by low temperature preannealing. OP: oxide precipitate, SF: stacking faults (ref. 37).

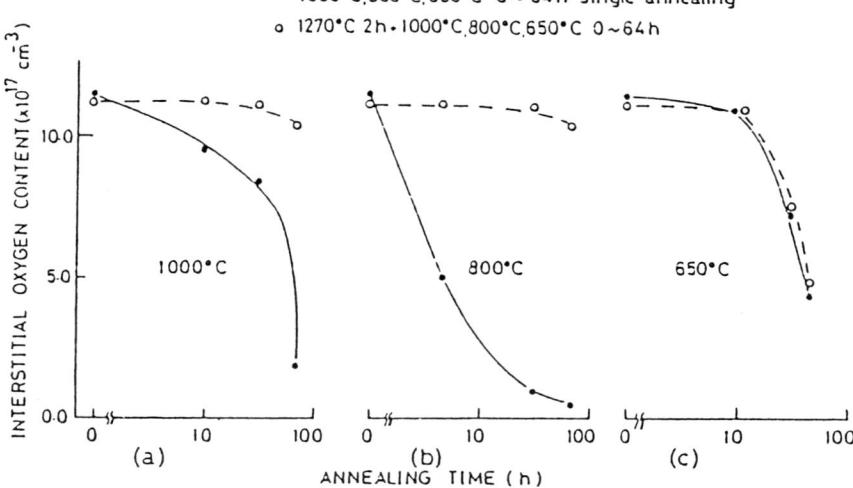

Fig. 9. Precipitation retardation by high temperature preannealing (ref. 3).

by Hu[23] and by Schaake et al.[33] In those cases, the method used was indirect, i.e. precipitated oxygen content measurement. Inoue et al.[2] measured the nucleation rate for up to a few hundred hours and found that delayed nucleation took place after several ten hours as shown in Fig. 10. The phenomenon was attributed to "time-lag in nucleation" which is well known

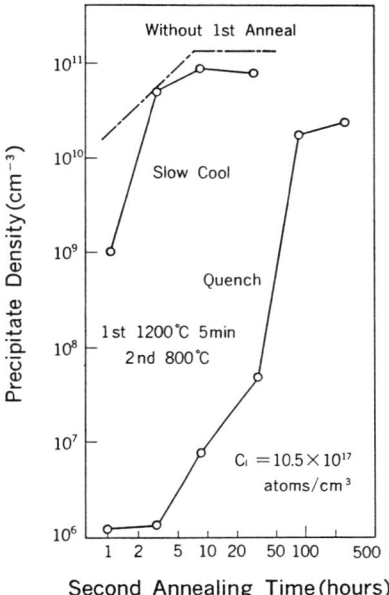

Fig. 10. Time-lag in nucleation in 850°C annealing due to high temperature preannealing (ref. 2).

as a characteristic of the homogeneous nucleation mechanism in the classical nucleation theory.[22] For high temperature annealing, especially in the undersaturation condition, a preexisting embryo size distribution is reduced down to the equilibrium distribution for high temperatures as shown in Fig. 6. Much time is required for the subsequent low temperature annealing to reform the equilibrium embryo size distribution through monoatomic reactions similar to those given in eqs. (8a) to (8d) in Section 5. Until reformation is completed, nucleation is not observed. Therefore, the time-lag in nucleation demonstrates the presence of nucleation, and shows that it is a characteristic of homogeneous nucleation rather than heterogeneous nucleation.

(4) *Growth fluctuation*

An inhomogeneous defect distribution is generally observed in as-grown or high temperature annealed specimens. Abe et al.[38] measured the oxygen striation through thermal donor activation and found that defects are generated on thermal oxidation in the regions with high oxygen concentrations. Nauka et al.[39] found that the oxygen concentration variation had an amplitude of up to 10% and suggested that an inhomogeneous as-grown oxide precipitate distribution is formed by the oxygen concentration variation. Wada et al.[29] found that about $10^6/cm^3$ precipitates are heterogeneously nucleated in as-grown crystal, and proposed that their origin might be some kind of growth fluctuation.

(5) *Dopant*

There have been many reports which demonstrated that the dopant species determines the defect species and density in crystals. De Kock et al.[27] first pointed out that crystals heavily doped with Sb contain a one order of magnitude lower precipitate density than other crystals. Precipitation enhancement was found in heavily boron-doped specimens.[40] In these studies, however, the oxygen content has not been determined by infrared measurement due to strong free carrier absorption. We have confirmed that the oxide precipitate nucleation rate is not different in heavily Sb doped crystal from that in other crystals as long as the oxygen content is equal.[41]

(6) *Annealing ambient*

Hu[42] reported that the oxidizing ambient strongly retards oxygen precipitation. Precipitation suppression took place throughout the entire bulk of the sample.

(7) *Wafer surface treatment*

Takano et al.[43] found that the precipitated oxygen content and precipitate density for annealing at 1000°C is enhanced in the specimen with backside damage.

Various effects described in items (5) to (7) have not yet been established. The authors listed above have discussed the effects in terms of intrinsic point defects whose type and concentration are generally considered to be affected by doping, annealing ambient, and surface treatment.

4.5 Growth mechanism

Yang et al.[18] have directly measured the growth of small octahedral precipitates formed by annealing at 1100°C using TEM. The precipitate growth follows nearly a parabolic relationship which is characteristic of diffusion limited growth. Wada et al.[17] have analyzed the growth of the most frequently observed platelet oxide precipitates in the range of 750–1050°C, and with sizes down to 100Å as shown in Fig. 11. This growth nearly follows a $t^{3/4}$ relationship. The standard equation for diffusion limited growth of precipitate (Johnson-Mehl equation) fits the data very well when the reported oxygen diffusion coefficient is used. As a result, precise oxide precipitate size control has become possible. Bourret et al.[19] observed smaller precipitates down to 20Å in size in specimens annealed at 650 and 870°C with preannealing at 450°C. These were also found to be platelet precipitates, but growth kinetics below 100Å is not yet clear.

4.6 Electrical and optical properties of precipitates

Yang et al.[18] suggested that oxide precipitates degraded minority carrier lifetime. Miyagi et al.[44] established the following; the density of large oxide precipitates (about 1000Å in size) in as-grown CZ silicon is closely related to the lifetime obtained by the non-contact photoconductive decay method as shown in Fig. 12. The precipitate density nucleated during annealing at 750 or 850°C shows the same tendency. It is found that precipitates larger than 100Å are electrically active.

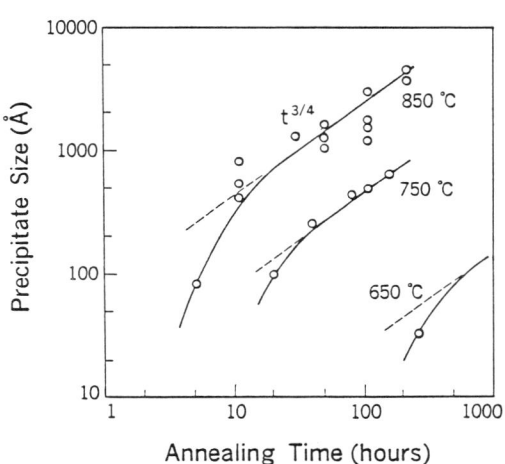

Fig. 11. Growth kinetics of oxide precipitates (ref. 17).

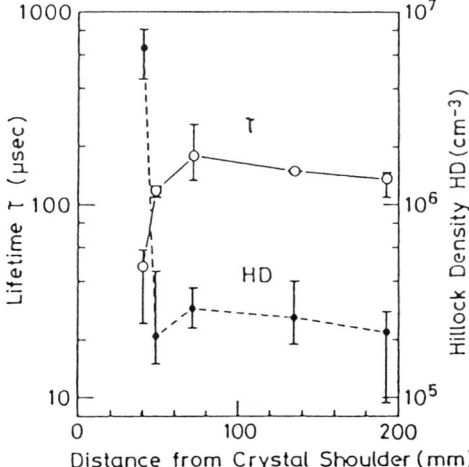

Fig. 12. Minority carrier lifetime as a function of precipitate density in as-grown crystal (ref. 44).

Kanamori et al.[45] found a new type of donors formed during prolonged annealing at 650°C where the thermal donors are quickly annihilated. The new donors have been suggested to be oxide precipitates,[46] but not yet been firmly established. Photoluminescence analysis is given in a separate article in these proceedings.[47]

4.7 Point defect formation due to oxygen precipitation

Oxygen precipitation results in the formation of secondary interstitial type dislocation loops. The silicon atomic concentration in oxide precipitates is nearly half of that in the silicon matrix. These facts led Booker et al.[48] to propose that silicon interstitials are emitted by oxygen precipitation. We have confirmed that number of silicon interstitials precipitated in the dislocation loops is about one-third of the precipitated oxygen atoms,[37] which supports the above consideration. However, the question of when and how point defects are formed is still a problem. It was found that oxygen precipitation at 750°C does not accompany dislocation loops, but precipitation at 850°C frequently does.[17] This suggests that nonequilibrium point defects are included at 750°C.

The formation of interstitial dislocation loops does not necessarily mean that the point defects responsible for controlling dislocation loop formation are silicon interstitials. It has been found that vacancy emission from dislocation loops rather than emission and diffusion of the interstitials from oxide precipitates control stacking fault growth. A detailed analysis of this phenomenon is given in these proceedings.[49]

5. Oxygen Thermal Donors

There are three types of known oxygen donors present in CZ silicon, thermal donors, new donors and deformation induced donors. Of these types, thermal donors have been the most intensively studied. However, neither their structure nor their formation kinetics have yet been firmly established (general kinetics information is found in a separate article in these proceedings[50]). Here, it is briefly pointed out that previously observed dopant dependence of thermal donor concentration have been explained by a model considering electronic interaction.

Thermal donors are formed by annealing at temperatures near 450°C. Their equilibrium density and initial formation rate are proportional to third and fourth powers of the oxygen content, respectively. Kaiser et al. applied a mass action law to the assumed monoatomic reaction for thermal donor formation kinetics and proposed a SiO_4 model.[32] This model can explain the above oxygen content dependence.[32] Thermal donor concentration has a tendency to increase in p^+ material and decrease in n^+ and carbon rich materials. Wada[51] gave attention to the dopant concentration dependence and added electron formation terms to the monoatomic reaction as

$$O_i + O_i = A_2^g + ge^-, \tag{8a}$$

$$A_2^g + ge^- + O_i = A_3^h + he^-, \tag{8b}$$

$$A_3^h + he^- + O_i = A_4^m + me^-, \tag{8c}$$

$$A_4^m + me^- + O_i = P_5. \tag{8d}$$

Here O_i is interstitial oxygen, A_i^k is a silicon-oxygen cluster including i oxygen atoms and with k-th valency, e is an electron, and P_5 is neutral oxide precipitates with 5 oxygen atoms. Thus the obtained formation kinetic formulae are

$$dN/dt = aD_i[O_i]^4 n^{-2}, \tag{9a}$$

$$N_{eq} = a/b[O_i]^3 n^{-2}. \tag{9b}$$

In eqs. (9a) and (9b), N is thermal donor concentration, a and b are constants, D_i is the oxygen diffusion coefficient, $[O_i]$ is the oxygen content, and n is electron concentration. The equation explains well the donor concentration dependence on dopant concentration as shown in Fig. 13.[51] In this figure, the effect of the electron environment on defect kinetics is first established.

The electrical as well as optical nature of thermal donors have also been intensively studied. Levels 0.07 and 0.15 eV below conduction band have been identified as thermal donor states,[52] and thermal donor divalency has been shown.[53] Deep level luminescence at 0.767 eV at 4.2 and 77 K has been related to thermal donors.[54]

New donors have been previously described in Section 4.6. Deformation-induced donors have been found in samples plastically deformed at 900°C.[55]

6 Defect Engineering

Oxygen related defects in silicon have recently been found to have beneficial effects on device performance. Thus "defect engineering" has been most successfully developed in silicon as summarized below.

6.1 Intrinsic gettering

The most successful "defect engineering" to date has been the gettering techniques.[56] Gettering utilizes crystallographic defects that form in electrically inactive regions of the wafer. These defects function as sinks for unwanted impurities which have sufficient mobility to diffuse to the defects.

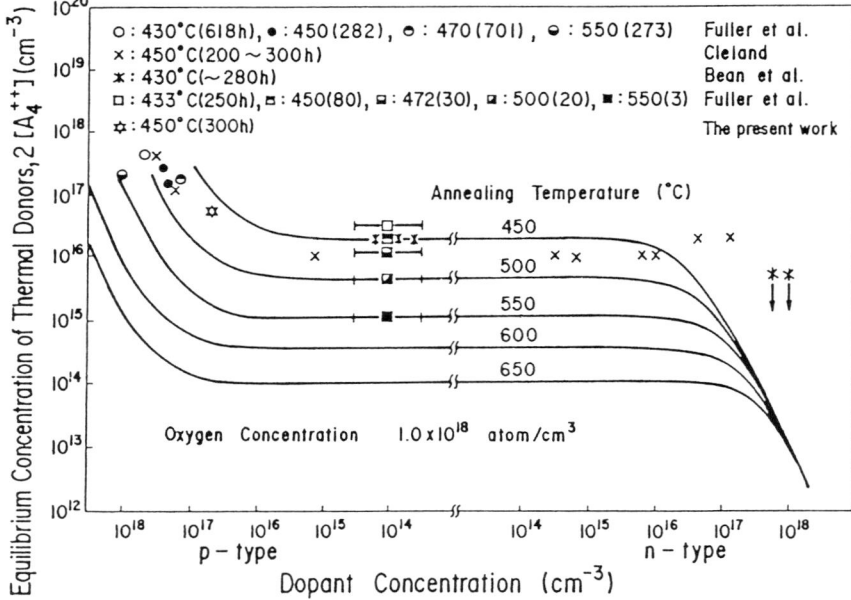

Fig. 13. Equilibrium oxygen thermal donor concentration dependence on dopant type and concentration (ref. 51).

Defects introduced mechanically at the wafer back-surface have been previously utilized.

Tan et al.[57] proposed an "intrinsic gettering" technique utilizing oxide precipitates as sources of prismatic dislocation loops capable of gettering metallic impurities. The effectiveness of this gettering has been demonstrated since the bipolar device yield improved considerably. Since this study was performed, there has been much research in this area, and now the technique is widely used. In addition to gettering metallic impurities, intrinsic gettering eliminates radiation damage due to reactive ion etching.[58]

Intrinsic gettering requires the simultaneous formation of surface denuded zone and bulk defect zone. Both of these zones depend on many factors, such as oxygen content, thermal behavior of a wafer and device fabrication process temperature. Many complicated annealing procedures are proposed to control these zones.[56] Furthermore, different wafer specifications must be used for different device structures and processes. Another problem is wafer warpage which may be due to heavy precipitation.

6.2 Mechanical strength

Silicon wafer warpage is induced in high temperature furnace operation. Warped wafers pose difficulties during photographic processes, and slip dislocations accompanying this warpage are associated with degraded electrical device properties. It has been recognized that oxygen-containing CZ silicon crystals are superior to FZ crystals in the mechanical response to thermal stress.[59] The mechanical strength of CZ silicon is attributed to resolved oxygen which might be resistant to dislocation movement under thermal stress. High oxygen content wafers, however, suffer from heavy precipitation which results in wafer warpage. Detailed analysis is given in these proceedings.[60]

6.3 Device application of oxygen donors

In as-grown p-type CZ silicon, thermal donor formation during the cooling period after crystal growth leads to overcompensation in high oxygen content regions. Based on this phenomenon, Kaiser and Keck[61] demonstrated p-n junction formation utilizing oxygen concentration variations introduced during float zoning in oxygen ambient followed by thermal donor activation using a 450°C heat treatment. Recently Chi et al.[62] have fabricated a multiple p-n junction on a longitudinal slice of CZ silicon which includes rotational oxygen striations. An increase in collection efficiency of this new solar cell structure over the conventional one has been demonstrated. Oxygen ion implantation has also been successfully utilized for conversion of the conductivity mode.[63] Its other application on full isolation by porous oxidized silicon (FIPOS) technique is also demonstrated. However, these two techniques are not free from either uncontrollable oxygen content variations due to growth or the process complexity resulting from ion implantation, respectively, and they don't seem as advantagoues as conventional dopant diffusion or implantation.

Recently, a novel oxygen donor incorporation method has been developed.[64] It utilizes laser annealing for localized selective donor incorporation, and requires essentially no additional processes, such as ion implantation and donor activation annealing. By this method, a p-n diode with good junction characteristics and MOSFET were successfully fabricated (Fig. 14). Oxygen content increase was confirmed in laser annealed region. Currently, the physical nature of the "laser induced donor" is being studied.

The above device application of oxygen donors utilizes electrical properties of defects, in contrast to the formerly established intrinsic gettering and wafer hardening which essentially utilize a mechanical property of defects. In this sense, the "donor device" is recognized as a second generation of device application of defects.

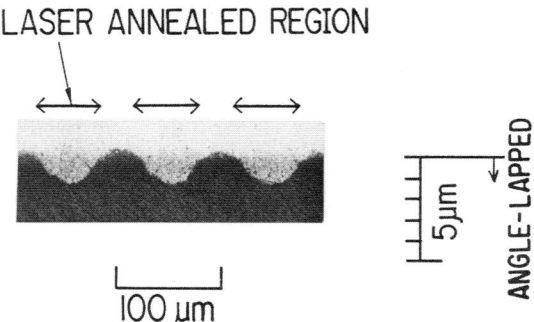

Fig. 14. Photograph of formed p-n junction.

7. Summary

Research on the fundamental properties of oxygen in silicon has been reviewed. Furthermore, the oxygen precipitation mechanism has been discussed in detail.

Defect engineering associated with oxygen in silicon is briefly summarized; intrinsic gettering using oxide precipitates, improvement in wafer mechanical strength by dissolved oxygen, and oxygen donor devices. The precise control of the oxygen content and its agglomerates including donors and precipitates is necessary for such device applications. From this viewpoint, more information must be obtained regarding oxygen behavior in silicon. In addition to the mechanical properties of oxygen and its agglomerates being mainly utilized, both electric and optical properties of oxygen should be tried for device application. As a result, further research must be conducted to better understand electric and optical properties of oxygen in silicon.

Acknowledgements

The authors would like to thank H. Hirata for discussions. We also thank H. Harada of Shin-etsu Handotai Co. for supplying heavily doped materials.

REFERENCES

1) J. R. Patel: *Semiconductor Silicon 1981*, ed. H. R. Huff *et al.* (The Electrochem. Soc., Princeton, 1981) p. 189.
2) N. Inoue, K. Wada, and J. Osaka: *ibid.* p. 282.
3) S. Kishino, Y. Matsushita, M. Kanamori, and T. Iizuka: Jpn. J. Appl. Phys. **21** (1982) 1.
4) H. Hirata and K. Hoshikawa: Jpn. J. Appl. Phys. **19** (1980) 1573.
5) K. Hoshikawa, H. Hirata, H. Nakanishi, and K. Ikuta: *Semiconductor Silicon 1981* p. 101.
6) T. Suzuki, N. Izawa, Y. Okubo, and K. Hoshi: *ibid.* p. 90.
7) Y. Yatsurugi, N. Akiyama, and Y. Endo: J. Electrochem. Soc. **120** (1973) 975.
8) W. Lin and D. W. Hill: J. Appl. Phys. **54** (1983) 1082.

9) G. Fiegl: Solid State Technology **26-8** (1983) 121.
10) H. J. Hrostowski and R. H. Kaiser: Phys. Rev. **107** (1957) 966.
11) T. Iizuka et al.: J. Electrochem. Soc. **132** (1985) 1707.
12) H. J. Hrostowski and R. H. Kaiser: J. Phys. Chem. Sol. **9** (1959) 214.
13) R. A. Craven: *Semiconductor Silicon 1981* p. 254.
14) Y. Takano and M. Maki: *Semiconductor Silicon 1973* p. 469.
15) J. C. Mikkelsen: Appl. Phys. Lett. **40** (1982) 336.
16) M. Stavola, J. R. Patel, L. C. Kimerling, and P. E. Freeland: Appl. Phys. Lett. **42** (1983) 73.
17) K. Wada, N. Inoue, and K. Kohra: J. Cryst. Growth **49** (1980) 749.
18) K. H. Yang, H. F. Kappert, and G. H. Schwuttke: Phys. Stat. Sol. **A50** (1978) 221.
19) A. Bourret, J. Thibault-Desseaux, and D. N. Seidmann: J. Appl. Phys. **55** (1984) 825.
20) P. E. Freeland, K. A. Jackson, C. W. Lowe, and J. R. Patel: Appl. Phys. Lett. **30** (1977) 31.
21) J. Osaka, N. Inoue, and K. Wada: Appl. Phys. Lett. **36** (1980) 288.
22) Ed. A. C. Zettlemoyer: *Nucleation* (Marcel Dekker, Inc. New York, 1969).
23) S. M. Hu: J. Appl. Phys. **52** (1981) 3974.
24) F. Shimura, H. Tsuya, and T. Kawamura: Appl. Phys. Lett. **37** (1980) 483.
25) G. S. Oehrlein, J. L. Lindstrom, and J. W. Corbett: Appl. Phys. Lett. **40** (1982) 241.
26) K. V. Ravi: J. Electrochem. Soc. **121** (1974) 1090.
27) A. J. R. de Kock and W. M. van de Wijgert: Appl. Phys. Lett. **38** (1981) 888.
28) N. Inoue, J. Osaka, and K. Wada: J. Electrochem. Soc. **129** (1982) 2780.
29) K. Wada, H. Nakanishi, H. Takaoka, and N. Inoue: J. Cryst. Growth **57** (1982) 1535.
30) C. S. Fuller, J. A. Ditzenberger, N. B. Hanney, and E. Buehler: Phys. Rev. **96** (1954) 833.
31) B. Pajot, H. Compain, J. Leronille, and B. Clerjaud: Physica **117-118B** (1983) 110.
32) W. Kaiser, H. L. Frisch, and H. Reiss: Phys. Rev. **112** (1958) 1546.
33) H. F. Schaake, S. C. Baber, and R. F. Pinizzoto: *Semiconductor Silicon 1981* p. 273.
34) C. Y. Yang, L. Forbes, and J. D. Peng: *Proc. Symp. Defects in Silicon* (Electrochem. Soc., Princeton 1983) p. 185.
35) M. Ogino: Appl. Phys. Lett. **41** (1982) 847.
36) S. Kishino et al.: J. Appl. Phys. **50** (1979) 8240.
37) H. Takaoka, J. Oosaka, and N. Inoue: JPN. J. Appl. Phys. Suppl. **18-1** (1979) 179.
38) T. Abe, K. Kikuchi, and S. Shirai: *Semiconductor Silicon 1977* p. 95.
39) K. Nauka, H. C. Gatos, and J. Lagowski: Appl. Phys. Lett. **43** (1983) 241.
40) G. A. Rozgonyi and R. J. Jaccodin: *Mat. Res. Soc. Symp. Proc.* (Elsevier Science Pub., 1983) Vol. 14, p. 181.
41) N. Inoue: unpublished.
42) S. M. Hu: Appl. Phys. Lett. **36** (1980) 561.
43) Y. Takano, H. Kozuka, M. Ogirima, and M. Maki: *Semiconductor Silicon 1981* p. 743.
44) M. Miyagi, K. Wada, J. Osaka, and N. Inoue: Appl. Phys. Lett. **40** (1982) 719.
45) A. Kanamori and M. Kanamori: J. Appl. Phys. **50** (1979) 8095.
46) A. Ohsawa et al.: J. Appl. Phys. **53** (1982) 5733.
47) M. Tajima: These proceedings.
48) G. R. Booker and W. J. Tunstall: Phil. Mag. **13** (1966) 71.
49) K. Wada and N. Inoue: These proceedings.
50) M. Suezawa: These proceedings.
51) K. Wada: Phys. Rev. B **30** (1984) 5884.
52) L. C. Kimerling and J. L. Benton: Appl. Phys. Lett. **39** (1981) 410.
53) M. Suezawa and K. Sumino: Materials Letters **2** (1983) 85.
54) M. Tajima, T. Matsui, T. Abe, and T. Iizuka: *Semiconductor Silicon 1981* p. 72.
55) M. Koguchi, I. Yonenaga, and K. Sumino: Jpn. J. Appl. Phys. **21** (1982) L411.
56) R. B. Swaroop: *Solid State Technology/July 1983* p. 97.
57) T. Y. Tan, E. E. Gardner, and W. K. Tice: Appl. Phys. Lett. **30** (1977) 175.

58) K. Ikuta, S. Nakajima, and N. Inoue: *Ext. Abs. 16-th Conf. Solid State Devices and Materials* (Jpn. Soc. Appl. Phys., Tokyo, 1984) p. 483.
59) S. M. Hu and W. J. Patrick: J. Appl. Phys. **46** (1975) 1869.
60) K. Sumino: These proceedings.
61) W. Kaiser and P. H. Keck: J. Appl. Phys. **28** (1957) 882.
62) J. Y. Chi, H. C. Gatos, and B. Y. Mayo: IEEE Trans. **ED27** (1980) 1306.
63) J. Y. Chi and R. P. Holmstrom: Appl. Phys. Lett. **40** (1982) 420.
64) Y. Mada and N. Inoue: Appl. Phys. Lett. **48** (1986)

ON THE FORMATION PROCESS OF THERMAL DONORS IN CZOCHRALSKI-GROWN SILICON CRYSTAL[1,2]

M. SUEZAWA

The Research Institute for Iron, Steel and Other Metals, Tohoku University, Sendai 980, Japan

1. Introduction

It is well known that donors are generated in silicon crystals supersaturated with oxygen due to heat treatment at temperatures around 450°C. They are termed thermal donors and abbreviated as TD in the following. They are annihilated due to heat treatment of the crystal above 500°C. Such generation and annihilation of TD seem to correspond to the development and dissolution of special kind of clusters of oxygen atoms, respectively.

Several models have so far been proposed on the structure of thermal donors. The so-called SiO_4 model due to Kaiser et al.[3] is based on the following experimental results.

1) The initial formation rate of thermal donors is proportional to the fourth power of initial concentration (abbreviated as O_I hereafter) of solute oxygen in the crystal.

2) The saturation concentration of thermal donors is proportional to the third power of O_I.

3) Many optical absorption lines are associated with thermal donors: These lines were first observed by Hrostowski and Kaiser[4] but were not classified.

They interpreted the above results under the assumptions that the dominant thermal donors were SiO_4, even though SiO_2 and SiO_3 might also be active as the thermal donors, and that oxygen clusters which included oxygen atoms more than four were electrically inactive.

The above results 1 and 2 were deduced from the measurements of the electrical resistivity at room temperature as the measure of the concentration of thermal donors. As already suggested in the above 3, many kinds of thermal donors are generated simultaneously due to the heat treatment of the crystal. We think that non-spectroscopic experimental methods such as the electrical resistivity measurements are valid only when all kinds of the thermal donors are generated at the same rates (or with the same dependence on the time of heat treatment). The formation process of each kind of thermal donors investigated with the technique of optical absorption is presented in the following.

2. Experimental Method

Specimens were prepared from n-type Czochralski-grown silicon crystals of different concentrations of oxygen doped with phosphorus at concentrations of about 1.5×10^{14} atoms/cm^3. The concentrations of solute oxygen of the crystals used are shown in Table 1.

Specimens were sealed into evacuated fused-quartz capsules and were subjected to heat treatment at 471.3°C for various durations resulting in the generation of thermal donors.

The concentrations of generated thermal donors were determined with the measurements of optical absorption at about 6 K.

3. Experimental Results

Figure 1 shows the optical absorption spectra due to thermal donors developed in silicon. Symbols nP_0, $nP_\pm$ mean to correspond to the optical transitions from 1s to nP_0, $nP_\pm$, respectively, in the effective-mass approximation. The optical absorption lines were classified as shown in Fig. 1 on the bases of the effective-mass approximation and also of the formation kinetics of thermal donors at various temperatures of heat treatment. Figure 1 shows the existence of six kinds of thermal donors (① through ⑥. We call them TD-1 through TD-6 hereafter.) between wavenumbers 550 and 375 cm^{-1}.

The strongest absorption line due to each kind of thermal donors corresponds to the $1s \rightarrow 2P_\pm$ transition. Therefore, we consider its intensity as the measure of the concentration of each kind of thermal donors.*

At first, Kaiser et al.'s assumption that the dominant thermal donor is SiO$_4$ or, in rather generalized expression, that the species of dominant thermal donor is invariant during heat treatment is examined.

Figure 2 shows the optical absorption spectra of specimen CN-3 which is heat-treated for various durations. A strong absorption line at about 520 cm^{-1} of the as-received state (curve 1) corresponds to the optical absorption due to the localized vibration of solute oxygen atoms. There are many absorption lines due to thermal donors on the curves 2 through 4. The strongest

---

*The concentration of each kind of thermal donors is proportional to the magnitude of optical absorption, or optical density, due to them, the proportional coefficient being not yet determined. Since the energy levels TD-1 through TD-6 are not very different from each other, the proportional coefficients of all kinds of thermal donors are thought to be almost the same.

Table 1.

Crystal	Oxygen concentration [atoms/cm^3]
CN-1	5.65×10^{17}
CN-2	6.39×10^{17}
CN-3	7.85×10^{17}

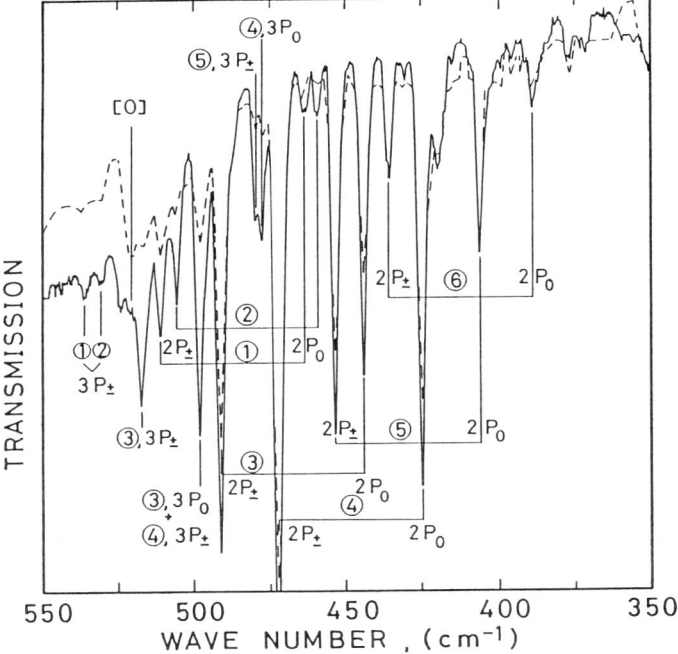

Fig. 1. Optical absorption spectra at Liq. He temperature for an initially p-type crystal (full line) and for an initially n-type crystal (dashed line).[1]

absorption line among them is that due to TD-3 after 20 min, TD-4 after 150 min and TD-5 after 1250 min durations of heat treatment. This shows that the species of dominant thermal donors changes with the duration of heat treatment. In other words, Kaiser *et al.*'s assumption (and also their conclusion) is not valid.

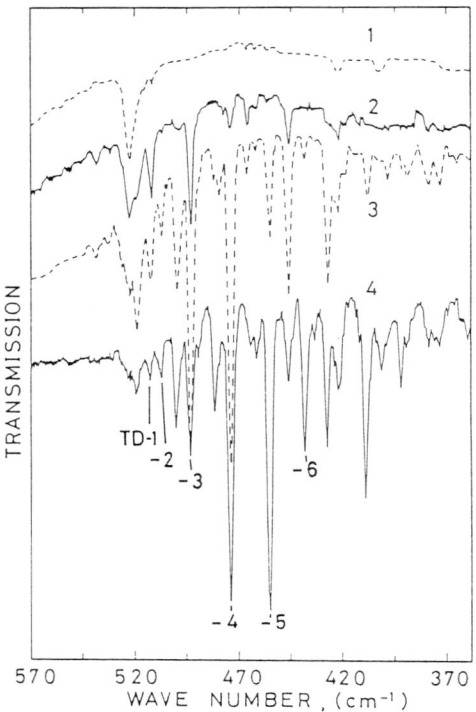

Fig. 2. Optical absorption spectra of crystals CN-3 annealed for various durations at 471.3°C.[2] The annealing duration t and the thickness of the specimens are as follows: (1) 0 min, 1.640 mm; (2) 20, 1.640; (3) 150, 1.640; (4) 1250, 0.397. The species of the donors which give rise to each absorption line are shown in the figure.

Figure 3 shows the dependence of optical absorption coefficients of various kinds of thermal donors on the duration of heat treatment. As the absorption coefficients (or concentrations) of TD-1 and TD-2 are small, we do not analyze their behavior in the following. The experimental data of TD-3 are not plotted only for the sake of simplicity. The characteristic of TD-3 is that the absorption coefficient (or concentration) is proportional to t, duration of heat treatment, for small t. The characteristics of TD-4, TD-5 and TD-6 are that their absorption coefficients are proportional to t^2 for small t. As pointed out with respect to Fig. 2, Fig. 3 also shows clearly that the species of thermal donors of the maximum concentration changes with duration of heat treatment.

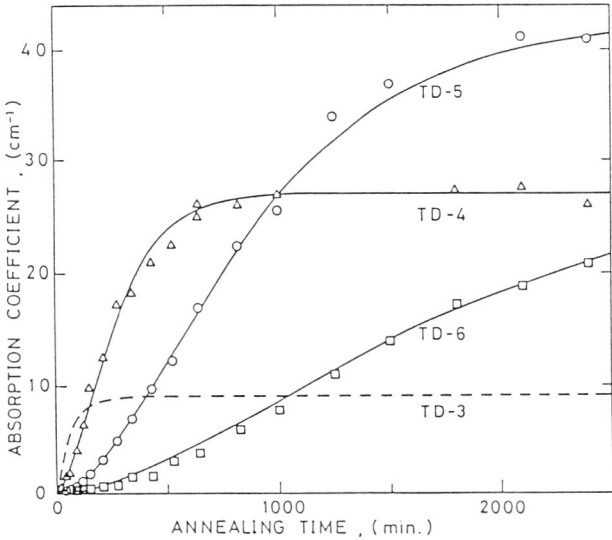

Fig. 3. The absorption coefficient at each line as a function of the duration of annealing at 471.3°C for crystal CN-3.[2]

4. Analysis and Discussion

In order to analyze the above results, Kaiser et al.'s analysis is extended to the present case as follows.

The formation process of clusters of oxygen atoms, which are probably the origin of thermal donors, is described with chemical rate eq. (1).

$$
\begin{aligned}
A_1 + A_1 &\underset{K_{-1}}{\overset{K_1}{\rightleftarrows}} A_2 \\
&\vdots \\
A_{n-1} + A_1 &\underset{K_{-(n-1)}}{\overset{K_{n-1}}{\rightleftarrows}} A_n \\
A_n + A_1 &\underset{K_{-n}}{\overset{K_n}{\rightleftarrows}} A_{n+1} \\
A_{n+1} + A_1 &\underset{K_{-(n+1)}}{\overset{K_{n+1}}{\rightleftarrows}} A_{n+2}, \\
&\vdots
\end{aligned}
\qquad (1)
$$

here A_n means the name of a cluster consisting of n oxygen atoms and also shows their concentration, K_n and K_{-n} the forward and backward reaction constants. A_1 at $t=0$ is the same to O_1 by definition.

It is implicitly assumed in eq. (1) that only the isolated oxygen atom can move in the crystal. Equation (1) is solved under the following three assumptions:
1) The reactions to form A_2 through A_n are in equilibrium.
2) The back reactions of A_{n+1}, A_{n+2}, ⋯ can be neglected.
3) A_1 can be regarded to be constant during heat treatment, i.e. $A_1 \equiv O_1$.

Following solutions of eq. (1) can be obtained under the above assumptions;

$$A_n = B_n \cdot A_1, \quad B_n = \frac{K_{n-1}}{K_{-(n-1)}} \cdots \cdots \frac{K_1}{K_{-1}} \quad (2)$$

$$A_{n+1} = \frac{K_n}{K_{n+1}} \cdot A_n \cdot [1 - e^{-K_{n+1}A_1 t}] \quad (3)$$

$$A_{n+2} = \frac{K_n}{K_{n+2}} \cdot A_n \cdot F(A_1 t)$$

$$F(A_1 t) = 1 + \frac{K_{n+1}}{K_{n+2} - K_{n+1}} e^{-K_{n+2}A_1 t}$$

$$- \frac{K_{n+2}}{K_{n+2} - K_{n+1}} e^{-K_{n+1}A_1 t} \quad (4)$$

Equations (3) and (4) show that eq. (5) hold for small t;

$$A_{n+1} \propto A_1^{n+1} \cdot t, \quad A_{n+2} \propto A_1^{n+2} \cdot t^2. \quad (5)$$

The full lines of Fig. 3 are the fitting curves of eq. (4) and the dotted lines of Fig. 3 is the fitting curve of eq. (3) with appropriate values of parameters. They fit to the experimental data in a fairly good manner.

Equation (5) shows the following. If the concentrations of generated thermal donors are proportional t or t^2 for small t, the dependence of proportional coefficient (abbreviated as K_e hereafter) on A_1 ($\equiv O_1$ according to assumption 3) reflect the number of oxygen atoms involved in each thermal donor.

Figure 4 shows the dependence of K_e on O_1. The numeral attached to each straight line shows its tangent. As suggested above, these numerals correspond to the numbers of oxygen atoms involved in various kinds of thermal donors. Namely, TD-3, TD-4, TD-5 and TD-6 include 5, 6, 7 and 8 oxygen atoms, respectively. The numbers of oxygen atoms included in TD-1 and TD-2 are not determined yet, but they are probably 3 and 4, respectively. S

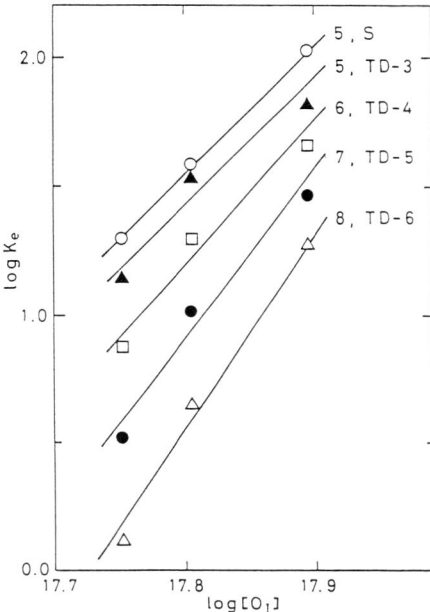

Fig. 4. The coefficients K_e of t and t^2 for small t against the initial concentration $[O_1]$ of dissolved oxygen atoms in logarithmic plots.[2]

shows the sum for TD-1 through TD-6 and, therefore, corresponds to the whole concentration of thermal donors. This should correspond to Kaiser *et al.*'s result, but the size of cluster determined from line S is 5 which is larger by 1 than theirs.

As shown above, it is clarified that thermal donors correspond to clusters of several oxygen atoms. Why do such clusters of oxygen atoms act as donors? Though this problem is not fully accounted for yet, we have the following picture.

The experimental result of optical absorption suggests that TD-1 through TD-6 are double donors.[5,6] The absorption spectra of Fig. 1 are well analyzed with the assumption that the symmetry of neutral thermal donors is tetrahedral (Td). The above results can be explained in a most feasible manner in terms of that the donor action is due to a substitutional oxygen atom. Isolated solute oxygen atoms occupy the interstitial sites. When several oxygen atoms form a cluster, one of them will replace a silicon atom and occupy the substitutional site. A substitutional oxygen atom forms covalent bonds with four silicon atoms of nearest neighbor. Four electrons of outer six electrons of oxygen atoms participate in the formation of covalent bonds.

Residual two electrons can be donor electrons. Neighbor oxygen atoms around a substitutional oxygen atom will occupy the interstitial sites and be electrically inactive forming covalent bonds among them and also with silicon atoms.

Finally, the author expresses his sincere thanks to Professor K. Sumino of Tohoku University for his fruitful discussions and encouragement, to Komatsu Electric Metals Co. for kindly supplying the crystals used in the investigation.

REFERENCES

1) M. Suezawa and K. Sumino: Mater. Letters **2** (1983) 65.
2) M. Suezawa and K. Sumino: Phys. Stat. Sol. **(a) 82** (1984) 235.
3) W. Kaiser, H. L. Frisch, and H. Reiss: Phys. Rev. **112** (1958) 1546.
4) H. J. Hrostowski and R. H. Kaiser: Phys. Rev. Lett. **1** (1958) 199.
5) A. R. Bean and R. C. Newman: J. Phys. Chem. Sol. **33** (1972) 255.
6) D. Wruck and P. Gaworzewski: Phys. Stat. Sol. **(a) 56** (1979) 557.

Defects and Properties of Semiconductors: Defect Engineering, edited by J. Chikawa,
K. Sumino, and K. Wada, pp. 227–259.
© KTK Scientific Publishers, Tokyo, 1987.

INTERACTION OF DISLOCATIONS WITH IMPURITIES IN SILICON

Koji SUMINO

The Research Institute for Iron, Steel and Other Metals, Tohoku University, Sendai 980, Japan

Abstract. A review is made of various aspects on dislocation-impurity interaction in silicon that have been clarified by the author's group. The subjects taken up are influence of impurities on the dislocation velocity, locking of dislocations by impurities, kinetics of impurity segregation on dislocations, effect of impurities on dislocation generation from flaws, effect of impurities on mechanical strength of silicon, strengthening of silicon wafers due to impurity doping, softening of silicon due to precipitation of impurities, and impurity state on dislocations.

1. Introduction

Many interesting problems are raised concerning the dislocation-impurity interaction in semiconductors from both fundamental and practical points of view. Some irregularities on the dislocation line such as kinks and jogs are now believed to be electrically active. Such electrically active parts of a dislocation naturally interact with electrically active impurities. The energy of the electrostatic interaction at an interatomic distance may be higher than that of the elastic interaction originating from the size misfit as far as the interaction between a dislocation and a single impurity atom is concerned. The electrostatic interaction may play a very important role in the phenomena occurring in a semiconductor crystal.

The atomic configuration at the dislocation core is deviated severely from that in the normal one and is never realized in a bulk crystal. It may be natural to think that the electronic structure of an impurity atom in the core region is changed from that in the bulk crystal. There may also be a possibility that some special chemical reaction proceeds at the core region of a dislocation that never takes place in the bulk region.

In practice, the dislocation-impurity interaction is commonly utilized to suppress the warpage of silicon wafers due to thermal stress at temperature cycling and also to getter harmful impurities from electroactive regions of

228 K. SUMINO

device elements. Thus, to understand the nature of dislocation-impurity interaction and to know how such interaction manifests itself due to various treatments of semiconductors are not only the subjects to be studied from scientific interests but also those to be clarified in order to find the combination of material and processing that results in the production of devices with high and reliable performance. This paper reviews several aspects on dislocation-impurity interaction in silicon that have been established by the author's group.

2. Influence of Impurities on the Mobility of Dislocations

Influence of various kinds of impurities on the dynamic behavior of dislocations in silicon has been investigated in a quantitative way with the technique of *in situ* X-ray topography by utilizing a high power X-ray generator.[1-4]

The velocity of dislocations in high purity silicon has been measured as a function of the shear stress τ and the temperature T and has been found to be expressed by the following empirical relation in the stress range 1–40 MN/m^2 and the temperature range 600–800°C;

$$v = v_0 \, \tau \, \exp(-E/k_B T), \tag{1}$$

where the magnitudes of v_0 are 1.0×10^4 and 3.5×10^4 m^3/MN·s and those of E 2.20 and 2.35 eV for 60° and screw dislocations, respectively, and k_B the Boltzmann constant.

Light element impurities dissolved in silicon that are electrically inactive are found not to affect the velocity of dislocations moving under high stresses. However, they influence the dynamic behavior under low stresses. Figure 1 shows the relations between the velocity of a 60° dislocation and the stress at various temperatures in silicon doped with light element impurities carbon, oxygen and nitrogen and also in high purity silicon. Dislocations are found to stop the motion when the stress is reduced lower than about 3 MN/m^2 if oxygen is dissolved at a concentration 15 ppm. The same is observed under stresses lower than about 4 MN/m^2 when silicon is doped with nitrogen at a concentration as low as 0.11 ppm. The velocity of dislocations in the oxygen-doped silicon is seen to decrease more rapidly than in the high purity silicon as the stress is decreased in the low stress range. Carbon at a concentration 2 ppm has no effect on the dislocation mobility.

Figure 2 shows the velocity versus stress relations for 60° dislocations at 647°C in silicon dissolving oxygen at a variety of concentrations. Vertical broken lines show the magnitudes of the critical stress for the stop of dislocation motion. It is seen that the deviation of the dislocation velocity from that in the high purity silicon in the low stress range and the critical stress

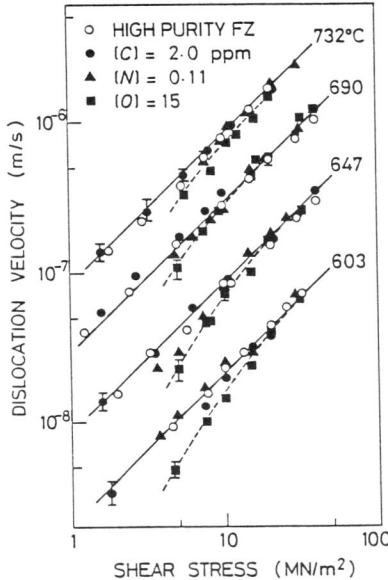

Fig. 1. Influence of light element impurities on the velocity versus stress relations at various temperatures of 60° dislocations. Open circles show the data for high purity silicon. (Imai and Sumino[2]).

for the stop of dislocations motion both increase as the concentration of dissolved oxygen increases.

The presence of donor impurities brings about an increase in the dislocation velocity from that in high purity silicon. The amount of the increase is determined only by the concentration and not influenced by the species of the impurities as seen in Fig. 3. The increase is related to the decreases in the magnitudes of both v_0 and E in eq. (1) as seen in Fig. 4. Even such donor impurities that enhance the dislocation velocity lead to the rapid decrease in the dislocation velocity with decreasing stress in the low stress range and also to the stop of dislocation motion under low stresses. Acceptor impurities in silicon show almost no influence on the dislocation velocity. Table 1 shows the critical stress of the stop of motion for initially moving dislocations in silicon crystals involving various kinds and concentrations of impurities.

In a summary, with a few exceptions, impurities in silicon generally interrupt the motion of dislocations that move at low velocities under low stresses. Electrically inactive impurities and acceptor impurities do not affect the velocity of dislocations moving under high stresses while donor impurities enhance the velocity. These hold for both 60° and screw dislocations.

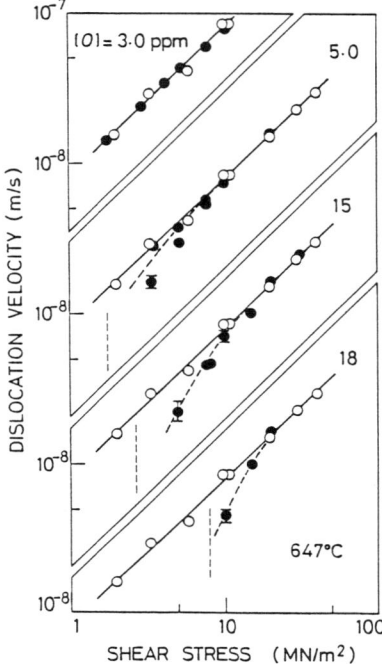

Fig. 2. The velocity versus stress relations at 647°C for 60° dislocations in silicon doped with oxygen at various concentrations. Open marks show the data for high purity silicon. (Imai and Sumino[2]).

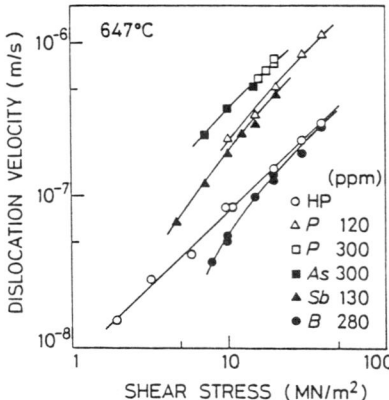

Fig. 3. Effects of electrically active impurities on the velocity versus stress relation at 647°C for 60° dislocations. Open circles show the data for high purity silicon. (Imai and Sumino[2]).

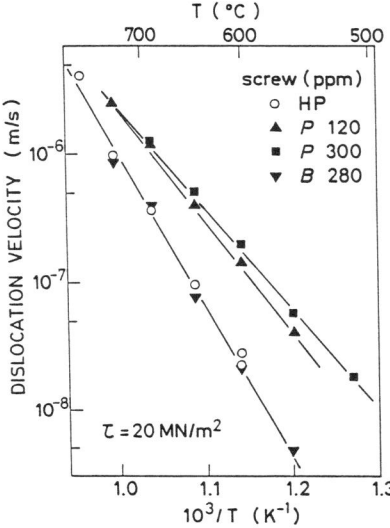

Fig. 4. Effects of electrically active impurities phosphorus and boron on the relation between the velocity and the reciprocal temperature $1/T$ for screw dislocations under a shear stress of 20 MN/m^2. Open circles show the data for high purity silicon. (Imai and Sumino[2]).

Table 1. Mean critical stress for stop of dislocation motion in silicon involving various kinds and concentrations of impurities. (Imai and Sumino[2]).

Major impurities	Concentration	Critical stress
B	0.00004 (ppm)	< 1.2 (MN/m^2)
C	2.0	< 1.2
N	0.11	4.2
P	240	8.5
O	3.0	< 1.2
O	5.0	1.8
O	15	3.0
O	18	8.0
{ P O	120 12	4.5
{ P O	300 12	11.5
{ B O	280 14	5.0

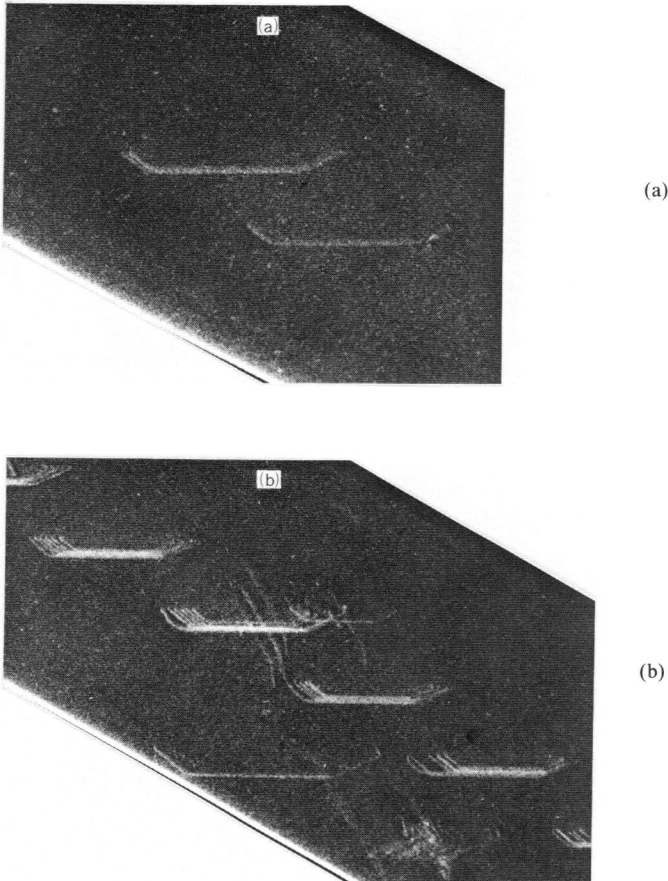

Fig. 5. Topographs of dislocation half-loops in motion at 647°C in silicon doped with oxygen at various concentrations (a) 3.0 ppm, (b) 5.0 ppm, (c) 15 ppm, and (d) 18 ppm. Except for (a), the topographs were taken under low stresses where deviations of the velocity from that in high purity silicon were remarkable. (Imai and Sumino[2]).

Irrespective of the purity of the crystal, the shape of a moving dislocation is a regular hexagon or a half-hexagon consisting of straight segments along <110> when its velocity is linear with respect to the stress. However, whenever a dislocation moves at a velocity that is not linear to the stress in the low stress range, the shape of the segments is perturbed from <110> straight lines and becomes irregular as seen in topographs in Fig. 5. Occurrence of such shape perturbation is reversible with respect to the change in the stress. The straightness of the segments is recovered if the stress is raised so that the

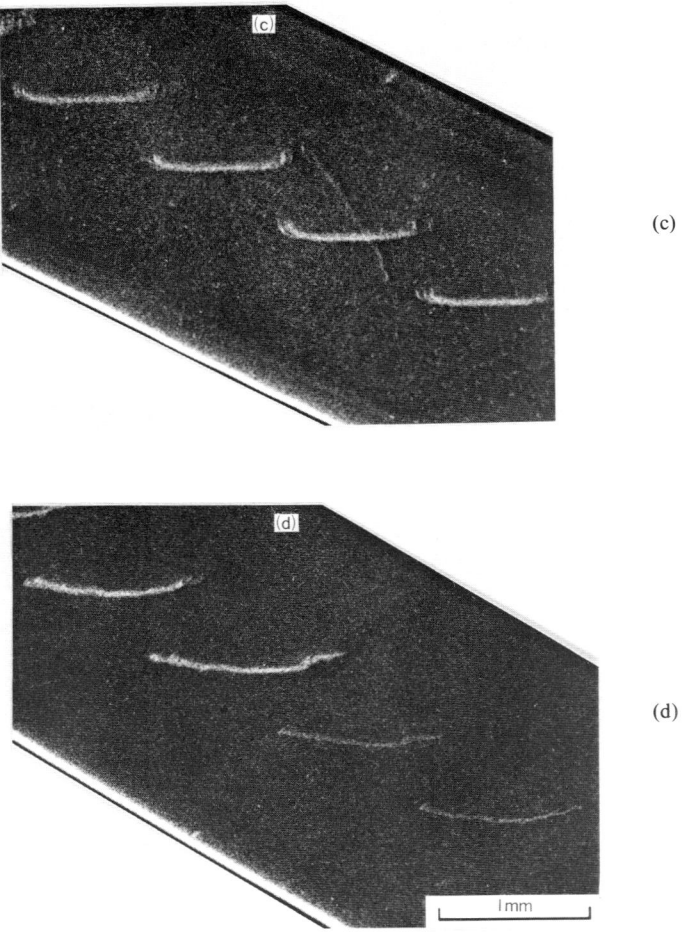

Fig. 5.

dislocation moves at a velocity that is linear to the stress. The perturbation in the shape of a moving dislocation starts to occur at a high stress in silicon of which the critical stress for the stop of dislocation motion is high. The shape perturbation and the stop of dislocation motion are both characteristic of impure silicon. They may be attributed to the interaction of dislocations with impurities. The former is thought to be caused by local locking of a slowly moving dislocation due to local development of impurity clusters on it, while the latter to the development of impurity clusters close along the dislocation line.

3. Impurity Locking of Dislocations

Impurity atoms are accumulated in the core regions of dislocations while the dislocations are kept at rest at elevated temperatures in impure silicon. The accumulation is caused by the interaction between dislocations and impurity atoms and results in the immobilization of dislocations. This effect is called impurity locking of dislocations and plays an important role in the strengthening of silicon. The impurity locking of dislocations is, of course, absent in high purity silicon and also is not practically observed in silicon involving impurities that interact weakly with dislocations. The external stress needed to start a locked dislocation is termed unlocking stress. The unlocking stress has been measured by means of *in situ* X-ray topography as a function of the temperature and the duration during which an initially fresh dislocation is kept under no applied stress at elevated temperature for silicon doped with impurities of various species and concentrations.[3,4]

The unlocking stress at 647°C in silicon doped with a variety of concentrations of oxygen is plotted against the duration of ageing of dislocations at the same temperature in Fig. 6. It increases with increasing duration of ageing and the increasing rate increases with increasing oxygen concentration. Figure 7 shows the data obtained for silicon doped with nitrogen or phosphorus in comparison with the data for oxygen-doped silicon. The locking of dislocations in silicon doped with 0.11 ppm nitrogen proceeds at almost the same rate as in silicon doped with 3 ppm oxygen and

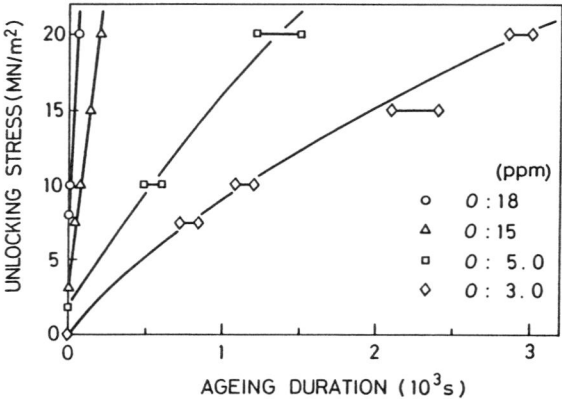

Fig. 6. Variation of the unlocking stress at 647°C for initially fresh dislocations against the duration of ageing at 647°C in silicon doped with oxygen at various concentrations. The concentrations of oxygen atoms are shown in the figure. (Sumino and Imai[3]).

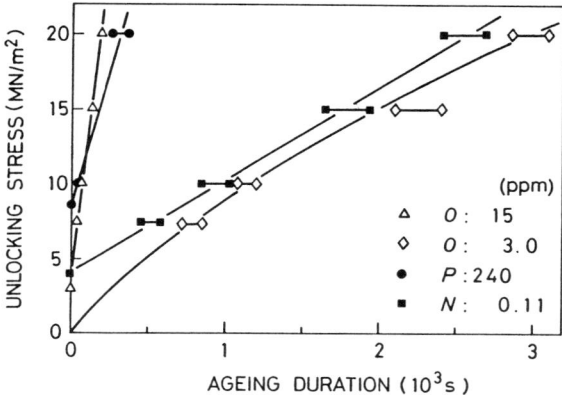

Fig. 7. Variation of the unlocking stress at 647°C for initially fresh dislocations against the duration of ageing at 647°C in silicon doped with various types of impurities. The species and concentration of impurities are shown in the figure. (Sumino and Imai[3]).

that in silicon doped with 240 ppm phosphorus at approximately the same rate as that in silicon doped with 15 ppm oxygen.

The locking strengths of individual atoms can be compared for various kinds of impurities with the numbers of impurity atoms accumulated on a unit length of a dislocation during the ageing that results in the same magnitude of the unlocking stress. The mean diffusion distance of individual impurity atoms during ageing is a good measure for the number of accumulated impurity atoms. A small diffusion distance means a small number of accumulated impurity atoms and, consequently, a high locking strength of individual impurity atoms. The comparison of the mean diffusion distances should be done for the same concentrations of dissolved impurity atoms since the higher the concentration of impurities in the matrix, the more the number of impurity atoms arrived at the dislocation during a given duration of ageing. The mean diffusion distance of an impurity atom is estimated from the diffusion constant and the ageing duration. In Fig. 8 the mean diffusion distances of nitrogen, oxygen and phosphorus atoms that are related to the unlocking stress of 20 MN/m^2 at 647°C are shown against the concentration in the matrix crystal. As is expected, the diffusion distance of oxygen atoms decreases approximately linearly with an increase in the concentration. Extrapolating the data for oxygen to the low and high concentration ranges, we know that the interactions of individual atoms of nitrogen and phosphorus with a dislocation are much stronger than that of oxygen.

The number of impurity atoms accumulated on a dislocation line can be calculated as a function of the temperature and the duration of ageing as will

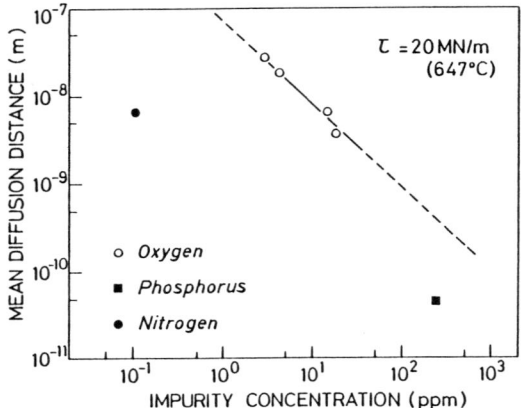

Fig. 8. Mean diffusion distances of various types of impurity atoms related to an unlocking stress of 20 MN/m² at 647°C plotted against the impurity concentration. (Sumino and Imai[3]).

be shown in a later section. Thus, the unlocking stress measured in the above is related to the number of impurity atoms accumulated on a unit length of dislocation. The relations obtained for silicon doped with various concentrations of impurity oxygen are shown in Fig. 9 in which the ordinate shows the unlocking stress and the abscissa the number of accumulated oxygen atoms N^* per unit length of dislocations. The relation is seen to be linear for any concentration of oxygen. While the relations are almost the same for silicon

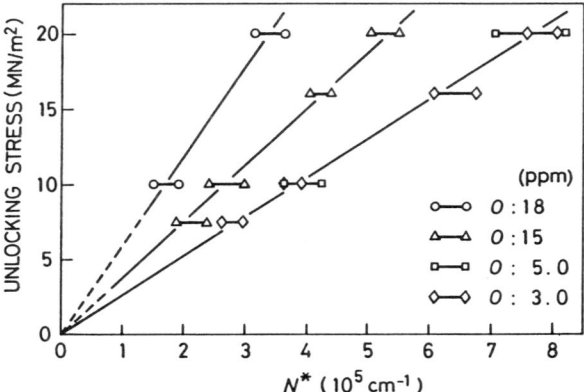

Fig. 9. Relation between the unlocking stress and the number N^* of oxygen atoms accumulated on a unit length of dislocations in silicon doped with oxygen at various concentrations. The concentrations of oxygen atoms are shown in the figure. (Sumino and Imai[3]).

crystals with rather low oxygen concentrations, an identical magnitude of the unlocking stress is achieved with a smaller number of accumulated oxygen atoms in a crystal highly concentrated with oxygen. This effect may be interpreted in terms of the inhomogeneity in the oxygen concentration within the crystal that becomes remarkable as the concentration of dissolved oxygen increases.

The magnitude of N^* seen on the abscissa of Fig. 9 shows that the accumulated oxygen atoms are distributed discretely along the dislocation line. Thus, we are led to a picture that some locking particles are arranged discretely along the dislocation line at an interval of $1/N$ where N is the density of the locking particles on the dislocation line. On the assumptions that the maximum energy of the interaction between a particle and a dislocation is E^* and that the interaction is of the short range nature, the theory of thermal activation gives the following expression for the releasing rate Γ of a dislocation from the locking particles:

$$\Gamma = LN \, v^* \exp[-(E^* - \tau \, b^2)/k_B T] \, , \qquad (2)$$

where, L is the length of the dislocation, v^* the frequency of the dislocation vibration, τ the stress and b the magnitude of the Burgers vector of the dislocation. Consequently, the unlocking stress τ_l is given by

$$\tau_l = N[E^* - k_B \, T \ln(LN \, v^*/\Gamma)]/b^2 \, . \qquad (3)$$

τ_l may be regarded to be linear with respect to N in accordance with the experimental relations in Fig. 9.

According to eq. (3), the theory predicts that the unlocking stress decreases linearly with an increase in the temperature T if the locking particles are of one kind, the slope being determined by N and the magnitude of the absolute zero temperature by $N \cdot E^*$. Figure 10 shows the experimental relations between the unlocking stress and the temperature for silicon crystals doped with oxygen at a concentration of 20 ppm and aged at 730°C for various durations between 5 and 30 min. In the temperature range above about 1050 K, the relations for all the crystals seem to be linear and parallel to each other. The relations suggest that the density of the locking particles along the dislocation line that control the releasing process in such temperature range is constant against the duration of ageing while the interaction energy increases with increasing duration of ageing. The changes in the magnitudes of E^* and N of the concerned particles with the duration of ageing are shown in Fig. 11. It is suggested that this kind of locking particles are developed on some special sites on dislocations of which density does not change with the ageing. The behavior of E^* seems to mean that the particles grow in size with an increase in the duration of ageing. Figure 10 shows that another linear

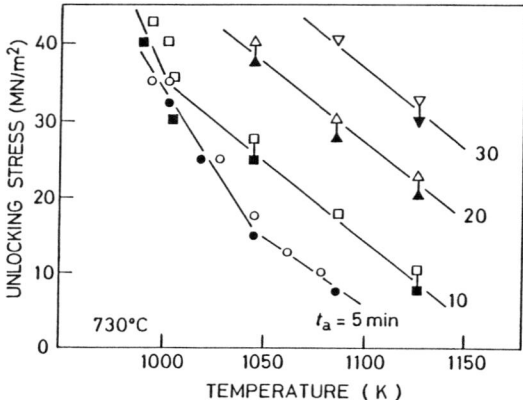

Fig. 10. Unlocking stress for dislocations in silicon doped with oxygen at a concentration of 20 ppm and aged at 730°C for various durations between 5 and 20 min plotted against the temperature. (Sato and Sumino[4]).

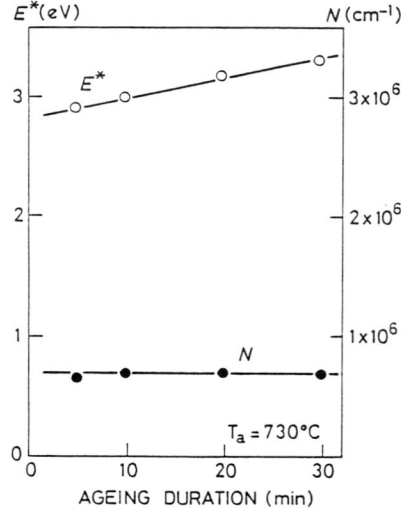

Fig. 11. Interaction energy E^* and the density N of locking particles along the dislocation line in silicon doped with oxygen at a concentration of 20 ppm plotted against the duration of ageing at 730°C. (Sato and Sumino[4]).

relation holds in the low temperature range in the crystal aged for short durations. This may mean that some other kind of locking particles also develop along the dislocation line in addition to those mentioned above. Evaluation of the number of accumulated oxygen atoms at the ageing suggests each locking particle to consist of several oxygen atoms. Such clusters of oxygen atoms on the dislocation line may be electrically active as donors as will be shown later. They may interact electrostatically with acceptor sites on dislocations that are possibly associated with jogs or kinks and result in an experimentally determined magnitude of interaction energy as high as about 3 eV.

4. Kinetics of Impurity Accumulation on a Dislocation

The kinetics of accumulation of impurity atoms on to a dislocation during ageing are investigated by means of computer simulation and the results are compared with the experimental data on impurity gettering by dislocations.

The rate of change in the impurity concentration c in the strain field of a dislocation is given by

$$\partial c / \partial t = D \nabla [\nabla c + (c/k_B T) \nabla E_i] , \qquad (4)$$

where t is the time, D the diffusion constant of the impurities and E_i the energy of interaction of an impurity atom with the dislocation. The interaction to be considered in the above equation is that of elastic origin since the electrostatic one may be, if any, of short range nature owing to the screening effect due to carriers that are abundant at temperatures concerned. The first term on the right hand side of eq. (4) is related to the diffusion flow originating from the inhomogeneity in the impurity concentration and the second term to the drift flow due to the interaction force between impurity atoms and the dislocation. If the dislocation is assumed to be straight along the z-axis, the problem is reduced to that in the two dimensions with the coordinate system (x, y). Assuming the misfit between the size of an impurity atom and that of the accommodating site is characterized by a parameter ε, E_i in eq. (4) is given by

$$E_i = (4/3)[(1 + v)/(1 - v)] \mu \, \varepsilon \, R^3 \, b \, \sin \alpha [y/(x^2 + y^2 + r_0^2)] , \qquad (5)$$

where v is the Poisson ratio, μ the shear modulus, α the angle between the dislocation line and the Burgers vector, r_0 the radius of the dislocation core. The dislocation is located at the origin of the coordinate.

The problem is treated in a numerical way with a high capacity computer[3,5)] on the assumptions that the initial distribution of the impurities is uniform throughout the crystal and also that the dislocation core is a perfect sink for impurity atoms arrived there. A dislocation is placed at the center of a two dimensional square lattice and the dislocation density in the crystal is defined to be the inverse of the area of the lattice. Further, the cyclic boundary condition is applied on the periphery of the lattice. The simulation shows that the concentration of impurity atoms in the dilatation field close to the dislocation core slightly increases in the early stage of ageing and, then, that in the region around the dislocation line starts to decrease. With increase in the duration of ageing, the diameter of impurity-depletion region expands outwards and the number of impurity atoms accumulated in the dislocation core increases steadily. In the late stage of ageing the impurities in the matrix are exhausted and the number of accumulated impurity atoms saturates.

The fraction f of the number of impurity atoms accumulated on the dislocations to the total number of impurity atoms in the crystal is expressed as a function of the duration of ageing t by

$$f = K t^n, \qquad (6)$$

where K is a constant that depends on the temperature and the interaction energy etc. The value of n for an early stage of ageing ($f < 5 \times 10^{-3}$) is independent of the dislocation density and is 0.79. Over a following wide range of f the value of n depends on the dislocation density. It increases with decreasing dislocation density and is 0.81, 0.87, and 0.89 for dislocation densities of 9.8×10^{10}, 6.1×10^{9}, and 1.6×10^{9} cm^{-2}, respectively.

The kinetics of the gettering of impurity oxygen by dislocations has been investigated with silicon doped with various concentrations of oxygen.[5] The crystals are deformed at an elevated temperature to introduce dislocations and, then, subjected to a high temperature annealing to eliminate deformation-induced defects other than dislocations and also to dissolve oxygen atoms segregated on defects during the deformation. Silicon crystals after such treatment involve dislocations with simple morphology. The density of dislocations can be controlled by adjusting the amount of deformation. The crystal is subjected to ageing at an elevated temperature to cause the accumulation of oxygen atoms on dislocations. The change in the concentration of oxygen atoms dissolved in the matrix crystal due to ageing is determined by the measurement of infrared absorption. The fraction f of oxygen atoms accumulated on dislocations is estimated on the assumption that all oxygen atoms removed from the matrix crystal are on dislocations. Figure 12 shows the change of f against the duration t of ageing at 900°C in silicon crystals with an oxygen concentration of 12 ppm. The numerals attached to curves are the dislocation densities in unit of cm^{-2}. Marks and solid lines show the experimental data and broken lines the results of calculation. The calculation shows a good agreement with the experiments both in the time law and in the magnitude. Thus, the simulation with eq. (4) is known to give reliable results.

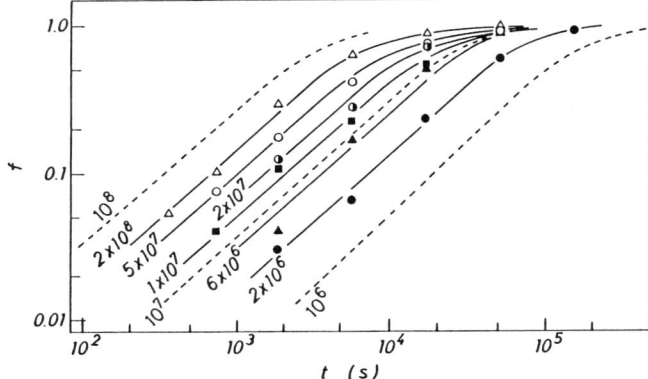

Fig. 12. Variation of the fraction f of oxygen atoms accumulated on dislocations in silicon doped with oxygen at a concentration of 12 ppm and involving dislocations at densities shown in the figure with the unit of cm^{-2} against the duration t of ageing at 900°C. (Yonenaga and Sumino[5]).

5. Effect of Impurities on Dislocation Generation

In an originally dislocation-free silicon crystal, dislocations are known to be generated from some irregularities at an elevated temperature under an applied stress far lower than the theoretical stress for the dislocation generation.[1] Such irregularities are present even on the surface prepared by chemical polishing and are thought to have atomic arrangements that facilitate the generation of dislocations under stress. In the case of a surface flaw made by scratching or indentation with a hard substance at room temperature, a thin layer with highly disturbed structure is introduced around it of which exact structure has not yet been fully understood. The structure of the disturbed layer recovers to that of the matrix when the crystal is heated to high temperature. Dislocations with various Burgers vectors that are favorable to release the strain caused by the displacement of materials associated with the flaw are introduced in this recovery process. If the crystal is under stress, the dislocations of the system that has the maximum Schmid factor move over a long distance and penetrate into the matrix crystal, and are macroscopically observed as the dislocation generation.[1]

Such dislocation generation from flaws takes place in high purity silicon over a wide stress range down to a very low stress. However, in impure silicon dislocations are not generated under low stresses. The critical stress for the dislocation generation increases with increasing impurity concentration and also with increasing temperature as shown in Fig. 13. This is due to the locking

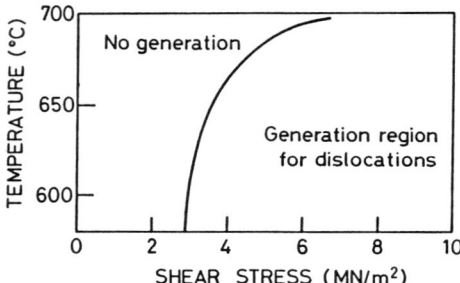

Fig. 13. Stress-temperature regions for generation and nongeneration of dislocations from Knoop indentations in silicon doped with oxygen at a concentration of 8 ppm. (Sumino and Harada[1]).

of dislocations by impurities. When the stress is low enough in impure silicon to allow the development of impurity clusters along dislocations as described earlier, the dislocations are immobilized and do not penetrate into the matrix crystal, resulting in the inhibition of dislocations generation.

6. Effect of Dislocation-Impurity Interaction on Mechanical Strength of Silicon

The presence of some kind of impurities has been known to bring about a remarkable increase in the mechanical strength of silicon under a certain condition. Such impurities are those having strong function for the locking of dislocations such as oxygen and nitrogen. Figure 14 shows stress-strain curves of dislocation-free crystals of silicon doped with various concentrations of oxygen[6] and that with nitrogen at 0.11 ppm[7] deformed at 900°C under a shear strain rate of $10^{-4} s^{-1}$. All crystals have optically smooth surfaces that are finished by chemical polishing. The yield strengths of the crystals show no systematic dependence on the concentration of the impurities and may be regarded to be approximately constant. On the other hand, a significant difference is seen among the mechanical strengths of initially dislocated crystals of silicon doped with the above kinds of impurities. Figure 15 shows stress-strain curves of annealed crystals of silicon doped with a variety of concentrations of oxygen that involve dislocations at densities approximately 10^6 cm^{-2} at 800°C. The yield stress is seen to be strongly dependent on the oxygen concentration and increases with increasing oxygen concentration. Doped nitrogen of 0.11 ppm has almost the same strengthening effect as that of oxygen of 3 ppm. To understand the characteristics of the above effect of impurities on the mechanical strength of silicon, we must first know the mechanism of yielding of silicon crystals.

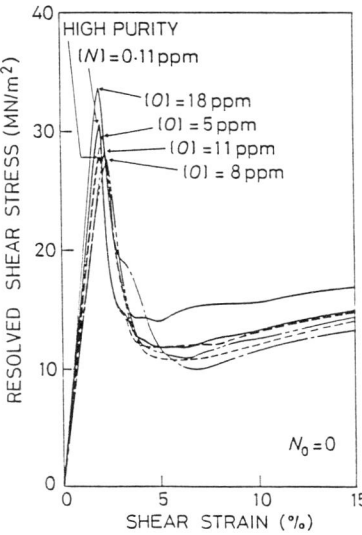

Fig. 14. Stress-strain curves of dislocation-free crystals of silicon of high purity and silicon doped with oxygen or nitrogen at concentrations shown in the figure obtained by tensile deformation at 900°C under a shear strain rate of 10^{-4} s^{-1}. (Yonenaga, Sumino, and Hoshi[6] and Sumino, Yonenaga, Imai, and Abe[7]).

The mechanical strength of silicon is characterized by the remarkable sensitivity to the temperature and the strain rate or stressing rate. Figures 16 and 17 show stress-strain curves of high purity silicon crystals in tensile deformation at various temperatures under a constant shear strain rate of 10^{-4} s^{-1} and those under various shear strain rates at 900°C, respectively.[8] All the crystals are dislocated at approximately the same densities of 2×10^4 cm^{-2} prior to deformation. It is seen in the figures that the mechanical strength of silicon decreases rapidly with increase in the temperature and increases markedly with increase in the strain rate. The stressing rate plays the same role as the strain rate does in determining the strength.

Now, the above characteristics in the mechanical strength of silicon crystals are interpreted in terms of dislocation dynamics. The yield point of a crystal is attained when the plastic strain rate $\dot{\varepsilon}$ reaches some critical value. $\dot{\varepsilon}$ is given by

$$\dot{\varepsilon} = N_D \, v \, b , \qquad (7)$$

where N_D is the density of dislocations that are moving in the crystal, v the velocity of dislocations and b the magnitude of Burgers vector of dislocations.

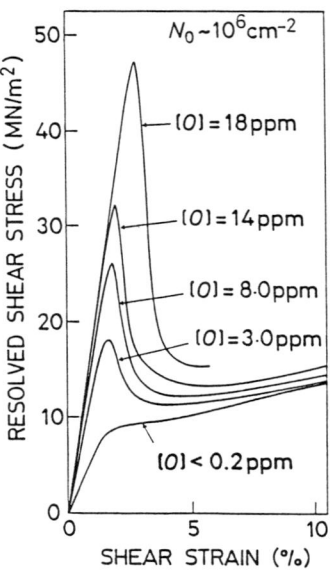

Fig. 15. Stress-strain curves of dislocated silicon crystals doped with oxygen at various concentrations shown in the figure obtained by tensile deformation at 800°C under a shear strain rate of 10^{-4} s^{-1}. The densities of initially involved dislocations are about 10^6 cm^{-2}. (Yonenaga, Sumino, and Hoshi[6]).

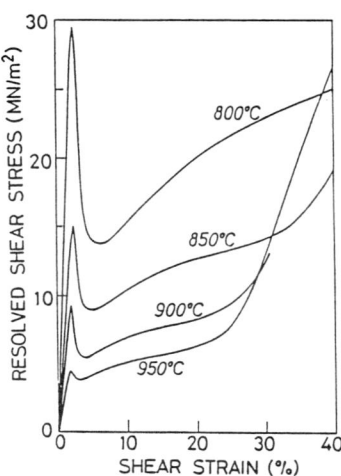

Fig. 16. Stress-strain curves of high purity silicon crystals with dislocations at a density of 2×10^4 cm^{-2} obtained by tensile deformation at various temperatures shown in the figure under a shear strain rate of 10^{-4} s^{-1}. (Yonenaga and Sumino[8]).

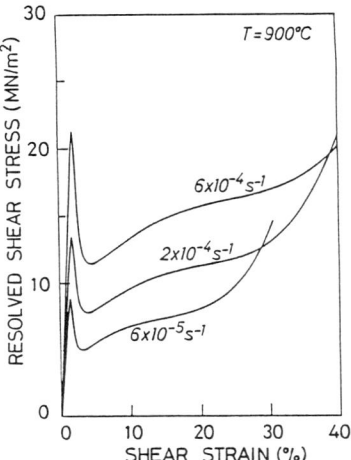

Fig. 17. Stress-strain curves of high purity silicon crystals with dislocations at a density of 2×10^4 cm^{-2} obtained by tensile deformation under various shear strain rates shown in the figure at 900°C. (Yonenaga and Sumino[8]).

The dislocation velocity v is given by eq. (1) as a function of the stress τ and the temperature T in high purity silicon.

N_D increases with time while the crystal is under stress due to self-multiplication of moving dislocations. The fundamental process of the multiplication is the spiral growth of the dislocation length around some irregularities on the dislocation such as super-jogs formed by some reason. The increasing rate $\dot{N}_D$ of the dislocation density may be taken to be proportional to the density of such multiplication centers on moving dislocations. As far as the strain range before the yield point is concerned, all dislocations in the crystal may be assumed to be moving. Then, $\dot{N}_D$ is taken to be proportional to N_D. Since the length of multiplied dislocations increases by means of motion of dislocations, $\dot{N}_D$ may be taken also to be proportional to the velocity v of dislocations moving in the crystal. Thus, $\dot{N}_D$ is given by

$$\dot{N}_D = \beta \, N_D \, v \,, \qquad (8)$$

where β is a coefficient. For the simplicity sake, the case that stress τ is increased at a constant rate γ with the time t is considered; namely,

$$\tau = \gamma \, t \,. \qquad (9)$$

With the use of eqs. (1), (8) and (9) and the assumption that β is constant, N_D is obtained as a function of the stress τ and temperature T as

$$N_D = N_0 \exp(k\tau^2)$$

$$k = (v_0 \beta / 2\gamma) \exp(-E/k_B T), \tag{10}$$

where N_0 is the density of mobile dislocations involved in the crystal before stress is applied. In the case of a dislocation-free crystal, N_0 may be taken to be proportional to the density of the irregularities that act as nucleation centers for dislocations when stress is applied. Figure 18 shows N_D as a function of τ for various temperatures for high purity silicon in which N_0 is taken to be 10^4 cm^{-2}. A broken line shows the line for $\dot{\varepsilon} = 10^{-4}$ s^{-1} and may be regarded as the yielding condition. Cross points of the curves give the magnitudes of the yield stress and those of the density of dislocations at the yield point. The mechanical strength of silicon is clearly seen to decrease rapidly with an increase in the temperature and also with a decrease in the stressing rate in agreement with experimentally observed facts.

There is another important parameter that affects the magnitude of the yield stress. Figure 19 shows the N_D versus τ relations at 800°C for high purity silicon for various values of initial density of dislocations (or the initial density

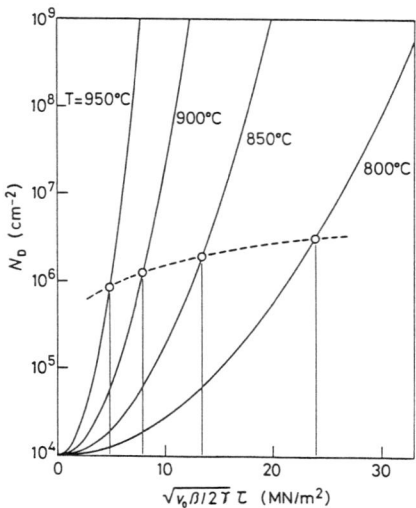

Fig. 18. Dislocation density N_D calculated for a high purity silicon crystal as a function of the stress τ at various temperatures shown in the figure. The crystal involves dislocations at a density of 10^4 cm^{-2} prior to application of the stress. Open circles show the yielding condition.

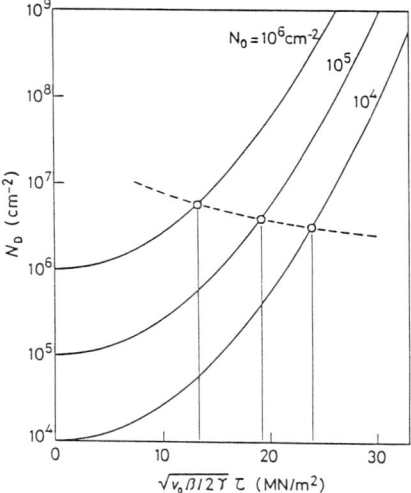

Fig. 19. Effect of the initial density N_0 of dislocations on the calculated N_D versus τ relation at 800°C for high purity silicon.

of dislocation sources). Again the cross points of the curves with a broken line give the mechanical strengths of the crystals. The strength of silicon is seen to increase significantly with a decrease in the density of dislocation sources involved prior to deformation according to the model mentioned above. This effect has indeed been observed in high purity silicon as shown in Fig. 20.

Backing to Fig. 15, we recognize that an increase in the oxygen concentration in the dislocated silicon crystals brings about the same effect as a decrease in the density of dislocation sources does. Impurity oxygen and nitrogen have already been seen to show strong locking effects on dislocations when the crystal is kept at elevated temperature under no applied stress. Locking of dislocations achieved at elevated temperatures may be so strong that most dislocations are not able to move even if a high stress is applied and, consequently, do not act as dislocation sources at deformation. Combining the above facts, we reach the idea that the strengthening of silicon due to doping of light element impurities working only in dislocated crystals is achieved by the locking of dislocations by the impurities. Figure 21 shows the yield stresses of silicon crystals doped with a variety of concentrations of oxygen plotted against the density of dislocations initially involved in the crystals.[6] The strength of dislocated silicon crystals depends rather sensitively on the dislocation density in spite of the fact that dislocations are thought to be locked firmly by impurity oxygen. When such crystals are stressed,

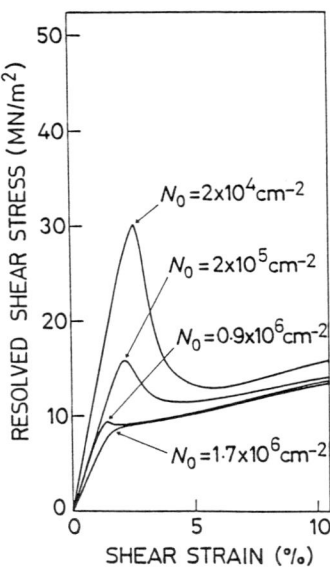

Fig. 20. Stress-strain curves of high purity silicon crystals involving dislocations at various densities shown in the figure obtained by tensile deformation at 800°C under a shear strain rate of 10^{-4} s^{-1}. N_0 shows the initial dislocation density. (Yonenaga and Sumino[8]).

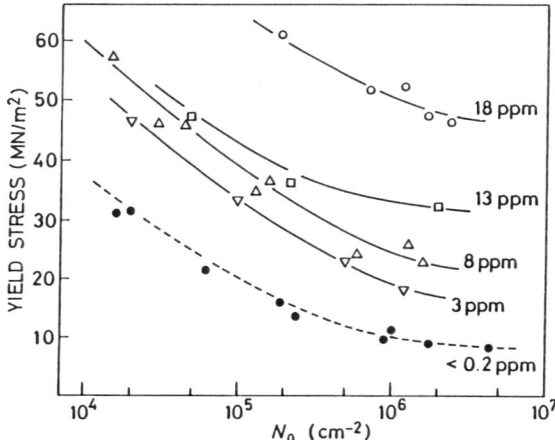

Fig. 21. Variation of the yield stress with the initial density of dislocations N_0 in silicon doped with oxygen at various concentrations shown in the figure. Yield stresses are for tensile deformation at 800°C under a shear strain rate of 10^{-4} s^{-1}. (Yonenaga, Sumino, and Hoshi[6]).

dislocations are observed to be working to some extent under stress before the yield point is reached. Thus, the yield stress is not determined by the releasing stress of locked dislocations from their impurity atmosphere. It may be concluded that some fraction of dislocations in the impure crystal remain not to be effectively locked by impurities even after annealing and that the fraction decreases with increasing concentration of impurities. Thus, a crystal with such locked dislocations behaves like a high purity crystal with a low density of dislocation sources and shows a high yield stress. The higher the concentration of the impurities in the crystal, the smaller the fraction of dislocations that can act as dislocation sources and, consequently, the higher the yield stress. Naturally, this type of mechanism of strengthening works only in dislocated crystals but not in dislocation-free crystals in agreement with the results in Figs. 14 and 15.

Impurity nitrogen at a concentration as low as 0.11 ppm works in dislocation locking in approximately the same way as oxygen at a concentration of 3 ppm does in silicon. Impurity carbon at concentrations 2 to 5 ppm is known to affect little the mechanical strength of silicon if the concentration of oxygen is lower than about 10 ppm. However, it shows a remarkable strengthening effect if the crystal involves impurity oxygen of which concentration is higher than the above value as shown in Figs. 22 and 23.[9] This seems to mean that carbon atoms in silicon plays a role that facilitates the segregation of oxygen atoms on dislocations.

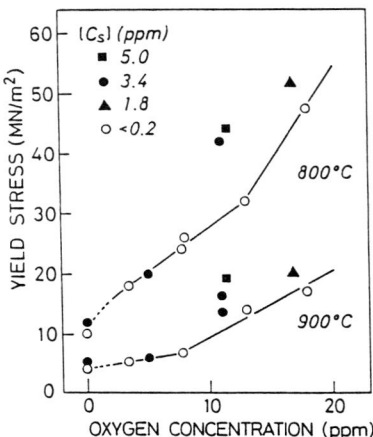

Fig. 22. Yield stress as a function of the oxygen concentration in silicon crystals with dislocation densities of about 10^6 cm^{-2}, the concentration of carbon [C_s] being taken as a parameter and shown in the figure. Open circles are the data for silicon doped with carbon at concentrations lower than the detectable limit. The yield stress is for tensile deformation at 800 or 900°C under a shear strain rate of 10^{-4} s^{-1}. (Yonenaga and Sumino[9]).

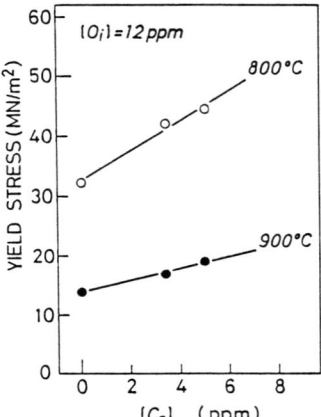

Fig. 23. Yield stresses at 800 and 900°C as a function of the concentration of carbon [C_s] in silicon with dislocations at densities about 10^6 cm^{-2} and with oxygen at concentrations about 12 ppm. Yield stresses are for tensile deformation under a shear strain rate of 10^{-4} s^{-1}. (Yonenaga and Sumino[9]).

7. Mechanism of Strengthening of Dislocation-Free Silicon Wafers by Doping of Impurities

In device production process it is well known that wafers of Czochralski-grown silicon show a much higher resistance to the occurrence of warping due to thermal stress compared with those of floating-zone-grown silicon. This phenomenon may be related to the strengthening of silicon due to the doping of oxygen. Warping of wafers becomes a more serious problem as the size of IC pattern becomes smaller and also as the wafer diameter becomes larger. Usually silicon wafers used for devices are cut from dislocation-free crystals. In this section, a mechanism is proposed on how originally dislocation-free silicon wafers are less susceptible to the occurrence of warping in thermal cycling when they are doped with impurities such as oxygen and nitrogen.

During device production process, silicon wafers are subjected to thermal cycling between room temperature and a high temperature repeatedly for many times. In the case that appreciable warp occurs, it becomes suddenly serious after some number of cyclings that depends on the profile of temperature ramping. Thermal stress is induced within a wafer during heating and cooling processes due to the inhomogeneity in temperature distribution, while the wafer is almost free from thermal stress while kept at a high temperature because of absence of the temperature inhomogeneity.

A silicon wafer naturally has some irregularities on the edge that act as preferential generation centers for dislocations. Dislocations are generated from such generation centers by thermal stress during heating and cooling processes in each thermal cycle. Generation centers of this kind exist in any wafers irrespective of the purity of the material. Hence, the generation of dislocations at the periphery of a wafer is thought always to take place at least on a microscopic scale. Probably, the penetration distance of dislocations from the periphery is smaller in silicon doped with impurities having a strong locking effect since the stress may decrease rapidly with the distance from the irregularity and dislocations in impure silicon stop moving when the stress becomes lower than a certain critical value. In absence of locking impurities, the generated dislocations move and undergo self-multiplication in succeeding thermal cycles when stress is applied during heating and cooling of the wafer. As the number of thermal cyclings increases, the density of dislocations increases steadily and the wafer becomes softer and softer by the reason shown in Fig. 19. After a certain number of cyclings, the wafer begins to yield and shows an appreciable warping. The situation is illustrated schematically in Fig. 24. On the other hand, in the presence of impurities that are effective in locking dislocations, most dislocations generated in the preceding cycle are locked while the wafer is kept at an elevated temperature under no stress in the thermal process. If the thermal stress during heating and cooling is controlled at such magnitude that the locked dislocations are not released from their impurity atmosphere, the density of dislocations can be kept low even after a number of thermal cyclings and, thus, the wafer retains a high mechanical

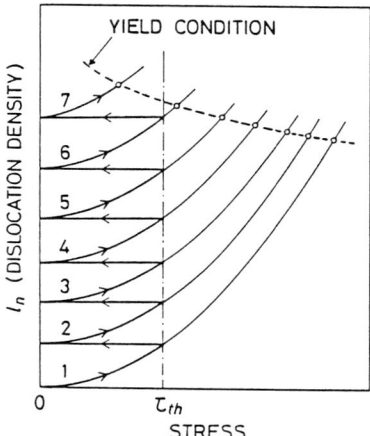

Fig. 24. Schematic illustration showing how the dislocation density increases with the number of cyclings accompanying a low stress in a wafer of high purity silicon.

strength. In this way, wafers of silicon doped with oxygen or nitrogen show a high resistance against the occurrence of warping due to thermal stress. Exactly speaking, doping of impurities that are effective to lock dislocations is not the technique to strengthen silicon material itself, but the technique to prevent wafers from becoming soft. The impurities having a strong interaction with dislocations and also having a high diffusion rate are effective agents for suppression of wafer warping.

8. Softening of Silicon due to Precipitation of Impurities

It is well recognized in the device production process that the high resistance against the warping in wafers of Czochralski-grown silicon is lost when the wafers are kept at temperatures around 1000°C for a rather long period. This phenomenon is related to the precipitation of super-saturated oxygen in Czochralski-grown silicon and can be studied in detail by ordinary mechanical tests.

Figure 25 shows how the stress-strain curve of an as-received Czochralski-silicon crystal, originally free from dislocations, changes with the duration of annealing at 1050°C.[10] The change in the characteristics in the stress-strain curve with the proceed of annealing is similar to that brought about by the increase in the initial density of dislocations in high purity silicon. Associated

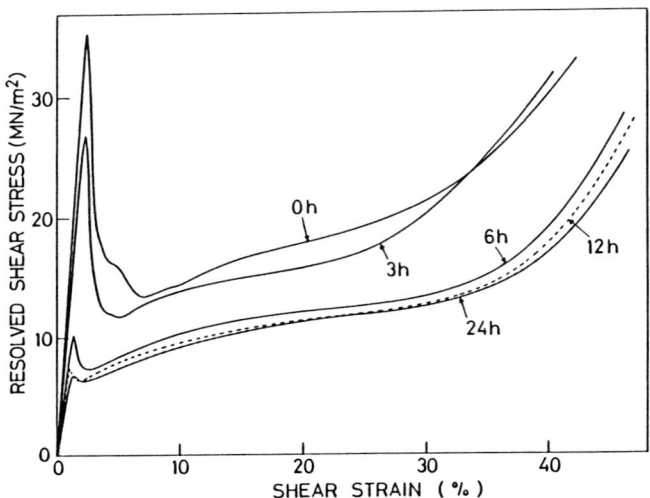

Fig. 25. Stress-strain curves of originally dislocation-free crystals of silicon doped with oxygen at a concentration of 20 ppm after subjection to annealing at 1050°C for various durations shown in the figure. The curves were obtained by tensile deformation at 900°C under a shear strain rate of 10^{-4} s^{-1}. (Yonenaga and Sumino[10]).

with such change in the mechanical behavior is the development of precipitates and various kinds of defects such as stacking faults of extrinsic type and loops of prismatic dislocations that are formed around precipitates. It may be concluded that the softening of Czochralski-silicon due to precipitation of oxygen is caused by the generation of defects that can act as effective dislocation sources under stress. The mechanical strength of the crystal prior to annealing at 1050°C is recovered when the softened crystal is annealed at temperature around 1300°C where the precipitates developed at 1050°C dissolve into the matrix. The precipitation-related defects are also eliminated by such high temperature annealing. The development of precipitates and related defects is strongly influenced by the thermal history of the crystal. If the crystal is first annealed at 1300°C, then cooled directly to 1050°C and held there, no appeciable development of both precipitates and defects takes place even after a prolonged annealing and the yield stress remains high. On the other hand, if the crystal is annealed at 1300°C and then once cooled to room temperature, the subsequent annealing at 1050°C leads to the development of numerous precipitates and defects and also to a drastic reduction in the yield stress. These facts imply that nuclei of precipitates or defects causing the softening of the crystal are formed during the cooling of the crystal from elevated temperature to room temperature and/or during heating from room temperature to 1050°C. It is known that, in general, a thermal treatment at a high temperature eliminates the nucleation sites for the precipitates that are destined to develop at lower temperatures.

The density of stacking faults developed during precipitation is confirmed to have no unique relation with the yield stress of the crystal. Thus, the stacking faults are not the dislocation sources that bring about the softening of the crystal associated with precipitation of oxygen. According to the analysis in terms of dislocation dynamics, precipitates of oxygen themselves are shown not to be the dislocation sources responsible for precipitation softening.[10] Prismatic loops of dislocations that are punched out from precipitates during the growth due to a large misfit strain are the most probable candidates for the dislocation sources concerned. In an early stage of precipitation process, precipitates formed are small in size and a small number of dislocations are punched out from each precipitate owing to a small misfit strain. The punched out loops are soon locked by oxygen atoms and are immobilized since oxygen atoms of a rather high concentration remain dissolved at this stage. As the precipitates grow with the duration of annealing, a large number of loops are punched out from each precipitate and, at the same time, exhaustion of dissolved oxygen in the matrix takes place. Thus, the loops are not effectively immobilized and drastic softening of the crystal occurs. This is the most plausible model that can describe the experimentally observed relation between the yield stress and the number of precipitated oxygen atoms seen in Fig. 26.[6] The difference in relations

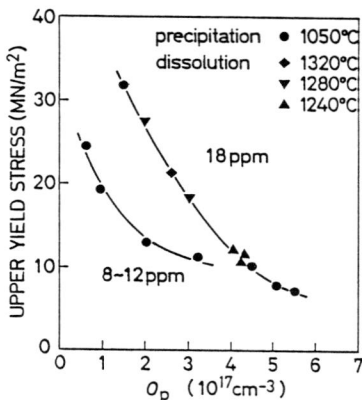

Fig. 26. Variation of the yield stress against the amount O_p of precipitated oxygen atoms for initially dislocation-free crystals of silicon doped with different concentrations of oxygen shown in the figure. The yield stress is for tensile deformation at 900°C under a shear strain rate of 10^{-4} s^{-1}. (Yonenaga, Sumino, and Hoshi[6]).

between highly and lowly concentrated crystals is neatly accounted for with the idea of locking of punched out dislocations by oxygen atoms remaining dissolved in the matrix.

The role of the concentration of impurity oxygen in the progress of precipitation softening is shown in Fig. 27,[6] in which the yield stresses of silicon crystals with various concentrations of oxygen are plotted against the duration of annealing at 1050°C. All the crystals are originally free from dislocations and are subjected to annealing at 1300°C followed by rapid cooling to homogenize the crystals. The data for silicon doped with nitrogen at 0.11 ppm is also shown for the purpose of comparison.[7] The yield stresses of all the crystals prior to subjection to annealing at 1050°C are almost the same as already seen in Fig. 14. Variation in the concentration O_i of dissolved oxygen atoms due to the annealing is also shown in the bottom of the figure. The decrease in the oxygen concentration due to the annealing has been shown to be related to the increase in the density of precipitates. Thus, the softening of the crystal is related closely to the increase in the density of precipitates and also to the decrease in the concentration of dissolved oxygen atoms. The concentration of dissolved oxygen atoms in silicon with an initial oxygen concentration of 18 ppm decreases rapidly in the early stage of annealing and becomes lower than those in silicon with lower initial concentrations of dissolved oxygen for the annealing duration longer than about 5 h. This seems to imply that the density of nuclei of precipitates depends strongly on the initial concentration of dissolved oxygen even if the crystal has been subjected to the homogenization treatment. This situation

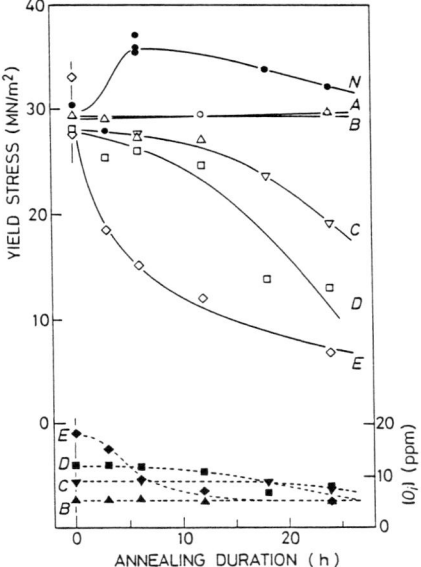

Fig. 27. Variation of the yield stress and the concentration $[O_i]$ of interstitially dissolved oxygen atoms against the duration of annealing at 1050°C for initially dislocation-free crystals of silicon doped with oxygen at various concentrations. A is for silicon with an oxygen concentration lower than 0.2 ppm, and B, C, D, and E are those with oxygen concentrations of 5.0, 8.0, 11, and 18 ppm, respectively. N is for silicon doped with nitrogen at a concentration of 0.11 ppm. The yield stress is for tensile deformation at 900°C under a shear strain rate of 10^{-4} s^{-1}. (Yonenaga, Sumino, and Hoshi[6] and Sumino, Yonenaga, Imai, and Abe[7]).

affects strongly the progress of precipitation softening in silicon. The yield stress of the nitrogen-doped silicon shows a peculiar behavior against the annealing at 1050°C. The exact origin for such behavior is not known at present. Precipitation or clustering process of nitrogen in silicon is an interesting problem to be clarified.

9. State of Impurities Accumulated on Dislocations

As pointed out in an early section, impurity atoms or clusters of impurity atoms located at the dislocation core may have the property which is different from that in the matrix crystal. This kind of information so far obtained for oxygen or nitrogen in silicon is reviewed in this section.[5,11-13]

As is well known, oxygen atoms interstitially dissolved in silicon give rise to an infrared absorption peak centered at a wave number of 1106 cm^{-1} that is associated with local mode of lattice vibration. On deforming a silicon crystal supersaturated with oxygen at 900°C, this absorption peak diminishes rapidly and, at the same time, a broad absorption peak is newly developed in the wave number range of 980–1080 cm^{-1} as shown in Fig. 28(a). The broad peak grows further upon annealing the deformed crystal at 900°C as seen in (b). When the deformed crystal is annealed at temperature as high as 1200°C for even a short duration, the broad absorption peak disappears completely and the height of the peak at 1106 cm^{-1} recovers to that prior to deformation. Deformation-induced defects other than dislocations are eliminated by this annealing at 1200°C. If the crystal is again annealed at 900°C, the height of the peak at 1106 cm^{-1} decreases gradually with the annealing duration but no absorption peak develops over the wave number range of 900–1300 cm^{-1} as shown in (c). Contrarily, when a dislocation-free crystal of the same oxygen concentration is annealed at 900°C, an absorption peak at 1225 cm^{-1} is observed to develop at the expense of the height of the peak at 1106 cm$^-$ as shown in (d). No definite peak is seen to develop in the wave number range of 980–1080 cm^{-1}. The peak at 1225 cm^{-1} is known to be due to the stable precipitate phase of SiO_2. It is concluded from the above observations that atomic arrangement around oxygen atoms segregated on dislocations or on deformation-induced defects

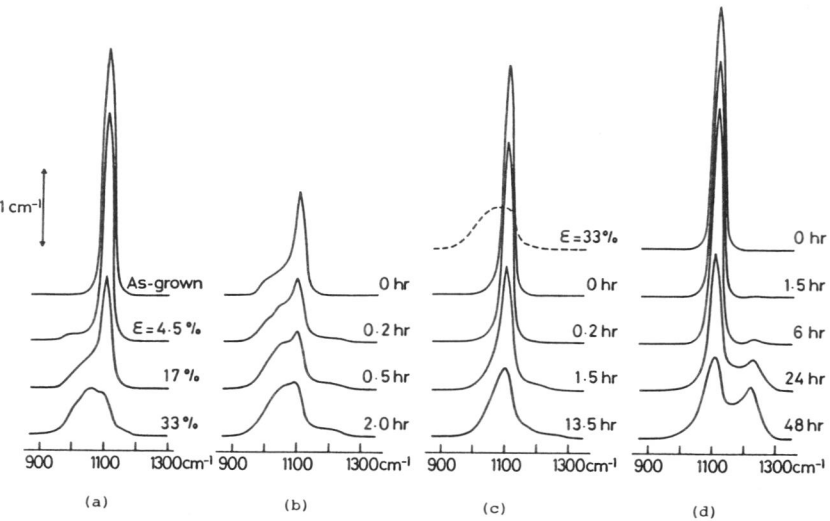

Fig. 28. Changes in infrared absorption spectra due to treatments at 900°C in silicon doped with oxygen at a concentration of 12 ppm. (a) Change due to deformation at 900°C. ε shows the shear strain. (b) Change due to annealing at 900°C after deformation of a shear strain of 17% at 900°C. The dislocation density is 2.0×10^8 cm^{-2}. Figures attached to the curves show the duration of annealing. (c) Change due to annealing at 900°C for the crystal involving only dislocations as deformation-induced defects at a density of 1.5×10^8 cm^{-2}. (d) Change in a dislocation-free crystal due to annealing at 900°C. (Yonenaga and Sumino[5]).

other than dislocations is much different from both that around an interstitial oxygen atom and that in a stable SiO_2 particle. Such atomic arrangement is fairly stable and does not transform to that in the SiO_2 phase even after a prolonged annealing at 900°C. The broad absorption peak in the wave number range 980–1080 cm^{-1} characteristic to the atomic arrangement developed on the defects other than dislocations is also observed to develop when so-called new donors are formed in a dislocation-free Czochralski-grown silicon due to annealing at about 650°C. Though it is not certain whether new donors themselves give rise to such an absorption peak or not, it is interesting to note that oxygen clusters developed on the defects at 900°C have an atomic arrangement similar to that develops in the matrix crystal at 650°C. Oxygen atoms accumulated on a dislocation is thought to be in the state much different from that of the above clusters. However, no clue is available at present to suppose the atomic bonding of such oxygen atoms because of the absence of the characteristic absorption peak.

It has been known that the free electron concentration in n-type silicon with a low concentration of oxygen decreases due to plastic deformation. When the amount of deformation exceeds a certain critical value which depends on the concentration of chemical donors, an originally n-type crystal is converted to a p-type crystal.[14,15] The experimental relation between the carrier concentration and the temperature is well described by the model that deep acceptors of which energy level is located at about 0.3–0.4 eV above the top of valence band are introduced by deformation. Such deformation-induced acceptors are ascribed to some irregularities on dislocations such as jogs and/or kinks or to defects other than dislocations such as clusters of point defects and/or various types of dislocation dipoles since their concentration is lower than that of geometrical dangling bonds arranged along the dislocation line by one or two orders of magnitudes. Contrary to the case of silicon of a low oxygen concentration, it is impossible to describe the carrier concentration versus temperature relation of a deformed crystal of n-type silicon doped with a high concentration of oxygen with the assumption that only deep acceptors are introduced by deformation. The fitting of a theoretical relation to an experimental one needs an assumption that relatively shallow donors are introduced due to deformation in addition to the deformation-induced deep acceptors.[11] Such donors are thought to be associated with oxygen atoms that are removed from the matrix crystal during deformation. The energy level of the donors is found to become deeper as the amount of deformation increases. It is worth noting that the donors are developed by deformation at temperature as high as 900°C where the generation of oxygen donors only by annealing is scarce in dislocation-free crystals of ordinary Czochralski-grown silicon.

It is questioned whether deformation-induced donors that develop in silicon doped with oxygen of high concentrations are related to oxygen clusters on dislocations or those on defects other than dislocations. Experiments have been conducted on oxygen-doped silicon crystals that involve only dislocations, deformation-induced defects other than dislocations being eliminated by annealing.[5] Figure 29 shows the concentration of donors generated in such crystals due to annealing at 900°C plotted against the concentration of oxygen atoms removed from the matrix crystals. It is known that some part of oxygen atoms segregated on dislocations act as donors since oxygen donors do not develop in dislocation-free regions at this temperature at any appreciable rate. It is interesting to note that the concentration of the generated donors is higher in the crystal with dislocations of a lower density. This seems to mean that there is some optimum concentration of segregated oxygen atoms along dislocations for the development of the donors.

Nitrogen atoms dissolved in silicon are also found to be gettered very effectively by dislocations at elevated temperatures. While nitrogen-related donors are developed in the matrix crystal due to annealing at temperatures around 750°C, nitrogen atoms accumulated on dislocations seem to be electrically inactive in contrast to the case of oxygen and have the function to kill the acceptor sites on dislocations.[13]

The phenomena appearing as results of dislocation-impurity interaction in semiconductors are rich in variety and extremely interesting from both fundamental and practical points of view. The author wishes to conclude this review with emphasizing that full understanding of the phenomena occurring in elemental semiconductors such as silicon offers the basis for clarifying the phenomena occurring in compound semiconductors that are more complex.

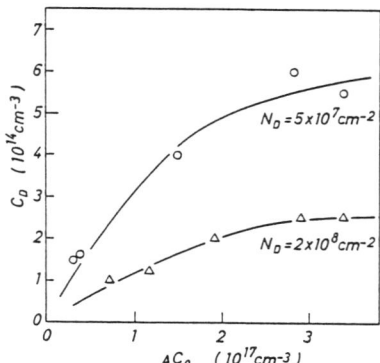

Fig. 29. The concentration C_D of generated donors due to annealing at 900°C plotted against the reduction ΔC_0 of interstitial oxygen atoms in silicon crystals involving dislocations at densities of 5×10^7 and 2×10^8 cm^{-2} and oxygen atoms at a concentration of 11.2 ppm. (Yonenaga and Sumino[5]).

The author is grateful to the members of his research group for their cooperation in the works on which the present review is based. Silicon crystals used in the investigations were offered by Hitachi Ltd., Komatsu Electronic Metals Co., Osaka Titanium Co., Shin-Etsu Handotai Co., and Sony Corp., to which the author wishes to express his gratitude. The works of the author's group have been supported by Grant-in-Aid for Scientific Research from the Ministry of Education, Science and Culture of Japan and also by Special Coordination Funds for Promoting Science and Technology from Science and Technology Agency of Japan.

REFERENCES

1) K. Sumino and H. Harada: Phil. Mag. **A44** (1981) 1319.
2) M. Imai and K. Sumino: Phil. Mag. **A47** (1983) 599.
3) K. Sumino and M. Imai: Phil. Mag. **A47** (1983) 753.
4) M. Sato and K. Sumino: *Proc. Yamada Conf. IX on Dislocations in Solids*, ed. H. Suzuki, T. Ninomiya, K. Sumino, and S. Takeuchi (University of Tokyo Press, Tokyo, 1985) p. 391.
5) I. Yonenaga and K. Sumino: *Proc. Yamada Conf. IX on Dislocations in Solids*, ed. H. Suzuki, T. Ninomiya, K. Sumino, and S. Takeuchi (University of Tokyo Press, Tokyo, 1985) p.385.
6) I. Yonenaga, K. Sumino, and K. Hoshi: J. Appl. Phys. **56** (1984) 2346.
7) K. Sumino, I. Yonenaga, M. Imai, and T. Abe: J. Appl. Phys. **54** (1983) 5016.
8) I. Yonenaga and K. Sumino: Phys. Stat. Sol. (a) **50** (1978) 685.
9) I. Yonenaga and K. Sumino: Jpn. J. Appl. Phys. **23** (1984) L590.
10) I. Yonenaga and K. Sumino: Jpn. J. Appl. Phys. **21** (1982) 47.
11) M. Koguchi, I. Yonenaga, and K. Sumino: Jpn. J. Appl. Phys. **21** (1982) L411.
12) K. Sumino: J. de Physique, Colloque C4 **44** (1983) C4-195.
13) H. Kamiyama and K. Sumino: *Proc. Yamada Conf. IX on Dislocations in Solids*, ed. H. Suzuki, T. Ninomiya, K. Sumino, and S. Takeuchi (University of Tokyo Press, Tokyo, 1985) p.399.
14) H. Ono and K. Sumino: Jpn. J. Appl. Phys. **19** (1980) L629.
15) H. Ono and K. Sumino: J. Appl. Phys. **54** (1983) 4426.

AUTHOR INDEX

Abe, T., 185

Chikawa, J., 143, 185

Fujimoto, I., 71

Harada, H., 185
Higuchi, H., 155

Inoue, N., 169, 197

Katoda, T., 111

Masui, T., 185
Mizuo, S., 155

Nishizawa, J., 25

Osaka, J., 197
Ozeki, M., 133

Sassa, K., 25
Shimanuki, Y., 25
Shinoyama, S., 87
Suezawa, M., 219
Sugano, T., 99
Sumino, K., 3, 227

Tajima, M., 37
Takikawa, M., 133
Tomizawa, K., 25

Wada, K., 169, 197

RAYMOND H. FOGLER LIBRARY

DATE DUE

BOOKS ARE SUBJECT TO
RECALL AFTER TWO WEEKS